Molecules and morphology in evolution: conflict or compromise?

Molecules and morphology in evolution: conflict or compromise?

EDITED BY

Colin Patterson

DEPARTMENT OF PALAEONTOLOGY
BRITISH MUSEUM (NATURAL HISTORY)
LONDON

The right of the University of Cambridge to print and sell all manner of books was granted by Henry VIII in 1534. The University has printed and published continuously since 1584.

CAMBRIDGE UNIVERSITY PRESS

CAMBRIDGE

LONDON NEW YORK NEW ROCHELLE

MELBOURNE SYDNEY

Published by the Press Syndicate of the University of Cambridge
The Pitt Building, Trumpington Street, Cambridge CB2 1RP
32 East 57th Street, New York, NY 10022, USA
10 Stamford Road, Oakleigh, Melbourne 3166, Australia

First published 1987

Printed in Great Britain by the Bath Press, Avon

British Library cataloguing in publication data

Molecules and morphology in evolution:
conflict or compromise?
1. Chemical evolution
I. Patterson, Colin
575 QH371

Library of Congress cataloguing in publication data

Molecules and morphology in evolution.
Papers presented at the Third International Congress
of Systematic and Evolutionary Biology, held at the
University of Sussex, 4–11 July 1985.
Includes bibliographies and index.
1. Evolution – Congresses. 2. Chemical evolution –
Congresses. I. Patterson, Colin. II. International
Congress of Systematic and Evolutionary Biology (3rd:
1985: University of Sussex)
QH359.M65 1987 575 86-23318

ISBN 0 521 32271 5 hard covers
ISBN 0 521 33860 3 paperback

BS

CONTENTS

CONTRIBUTORS

J. E. Ahlquist
Irvine Hall, Ohio University,
Athens, Ohio 45701, USA

P. J. Andrews
Department of Palaeontology
British Museum (Natural History)
Cromwell Road
London SW7 5BD, UK

W. R. Atchley
Department of Genetics
University of Wisconsin
Madison, Wisconsin 53706, USA

M. J. Bishop
Department of Zoology
University of Cambridge
Downing Street
Cambridge CB2 3EJ, UK

J. Czelusniak
Department of Anatomy
Wayne State University School of Medicine
540 East Canfield Avenue
Detroit, Michigan 48201, USA

W. M. Fitch
Molecular Biology
Ahmanson Center for Biological Research
University of Southern California
Los Angeles, California 90089-1481, USA

A. E. Friday
Department of Zoology
University of Cambridge
Downing Street
Cambridge CB2 3EJ, UK

M. Goodman
Department of Anatomy
Wayne State University School of Medicine
540 East Canfield Avenue
Detroit, Michigan 48201, USA

M. C. McKenna
Department of Vertebrate Paleontology
American Museum of Natural History
Central Park West at 79th Street
New York, NY 10024, USA

M. M. Miyamoto
Department of Zoology
University of Florida
Gainesville, Florida 32611, USA

C. Patterson
Department of Palaeontology
British Museum (Natural History)
Cromwell Road
London SW7 5BD, UK

C. G. Sibley
Tiburon Center for Environmental Studies
San Francisco State University
Box 855, Tiburon, California 94920, USA

C. R. Woese
Department of Genetics and Development
University of Illinois at Urbana-Champaign
515 Morrill Hall
505 South Goodwin Avenue
Urbana, Illinois 61801, USA

PREFACE

This book is one result of the Third International Congress of Systematic and Evolutionary Biology, held at the University of Sussex, 4–11 July 1985. These congresses, one every 5 years, are occasions for interdisciplinary exchange, when specialists try 'to enlighten the uninitiated and to interest and persuade the sceptic', a quote from a Congress document which is not a bad description of our aims in the book: we hope it will be accessible to undergraduates (*someone* has to be uninitiated), and persuasive to sceptics of any age or kind.

In proposing 'Molecules versus morphology' as a symposium topic at the Sussex Congress, the model I had in mind was the conflict over human evolution between molecular phylogeneticists on the one hand, and morphologists and palaeontologists on the other. In textbooks hardly more than a decade old, it is common to find morphology-based diagrams placing the split between the ape and human lineages in early Miocene or late Oligocene times, 20–30 MyrBP (million years before the present), whereas partisans of the molecular approach were giving estimates of only about 5 MyrBP for the ape–human split; and there were advocates of several different patterns of relationship between apes and humans. These differences were not merely due to emphasis on different lines of evidence – fossils versus antigens, perhaps – because different approaches to classification were also involved. Traditionally, apes and humans have been placed in different families: Pongidae for apes, Hominidae for us. Yet more than 20 years ago, Morris Goodman, one of our contributors, was arguing (on the basis of serum proteins) that Pongidae should contain only *Pongo*, the orang utan, and that chimpanzees and gorillas should join us in Hominidae. Those disputes, if not entirely settled, are at least a great deal more clearly focused now, and in Chapter 2 Peter Andrews summarises the current state of play in hominoids. As a morphologist who has come to grips with the molecular evidence, he was the obvious choice for hominoids, and I planned that the meeting and the book should move outwards from there, from animals that we all know into wider and less familiar circles. So I sought similarly qualified people to cover mammals as a whole, birds, other vertebrates, invertebrates, and single-celled life. Plants had to be omitted, principally for lack of space.

To my great pleasure (and surprise), all my first-choice authors agreed to contribute, and with one tragic exception, all are here in the book. The exception is Tom Schopf, a polymath palaeontologist at the University of Chicago who, though past 40, had just taken a sabbatical year to learn DNA sequencing at CalTech. Tom proposed to cover invertebrates, concentrating on the interrelationships of phyla in the area of the supposed split between protostomes (annelids, molluscs, arthropods, etc.) and deuterostomes (echinoderms, chordates, etc.). But Tom Schopf died suddenly, and distressingly young, in 1984 (see obituary by Stephen J. Gould in the journal that Tom founded, *Paleobiology*, **10**, 280–5). His absence leaves a gap in our coverage between the vertebrates, which are explored thoroughly in Chapters 2–6, and prokaryotes, covered by Carl Woese in Chapter 7. (Many biologists imagine that discovery of new species or even new génera is success; Carl Woese discovered a new superkingdom, the Archaebacteria, equal in rank to eukaryotes [Fig. 7.1], and in the nature of things, no one is likely to have that experience again.) This gap may not matter too much; our aim is to compare morphology and molecules as guides to phylogeny, and it is also in the nature of things that phylogeny, the history of life, is more thoroughly studied the closer one gets to our own species. That anthropocentric bias is reflected in the structure of the book. But even among anthropoids, phylogeny is hypothesis rather than fact, or reconstruction rather than documentation. In evaluating different approaches to phylogeny, what we lack is a known phylogeny that might provide an objective test of those approaches. In the last chapter, Walter Fitch (one of the founders of molecular phylogenetics) and Bill Atchley provide just that, a documented phylogeny, from the unexpected source of laboratory mice. Surprisingly, the history of inbred laboratory strains, though measured in decades rather than millenia or more, has generated enough differentiation for a phylogeny to be retrievable from both molecular and morphological data. Perhaps less surprisingly, the phylogenies differ, but one does match known history. Is this a triumph for that technique, or simply a reflection of the relatively crude data used for the alternative?

To give the topic of the book a setting, I have written an introductory essay setting out my own perception of the concepts behind the development of morphological and molecular approaches to phylogeny reconstruction. I hope it is not too biased by ignorance or prejudice.

As editor, I am indebted first to all the contributors, for agreeing to take part, for (mostly) meeting or even anticipating deadlines, and for tolerating my nit-picking through their manuscripts. I am particularly grateful to those contributors who came to the meeting at Sussex (largely at their own expense) and gave papers: Peter Andrews, Malcolm McKenna, Charles Sibley, Adrian Friday, Mike Miyamoto and Walter Fitch. At Cambridge University Press, I should like to thank Robin Pellew for his interest and support, and Rebecca Freeman for her meticulous copy-editing.

Colin Patterson

1 Introduction

Colin Patterson

To retrieve the history of life, to reconstruct the evolutionary tree, is still the central aim of evolutionary biology. This book is about the theory and practice of meeting that aim, and it brings together some of the newest and oldest parts of science. The new part is that branch of molecular biology concerned with comparing DNA or its protein products from different species. The traditional part is comparative morphology, a discipline with a history going back to antiquity. The reason for bringing molecular phylogenetics and comparative morphology together in the 1980s is to evaluate their performance: have molecules superseded morphology as guides to the history of life, or are the two approaches sides of the same coin, with the same problems and limitations? Do molecules and morphology give the same picture of the history of life, or two more or less distorted views of the same picture, or two quite different pictures?

From morphology to molecules

When Darwin published *On the Origin of Species*, in 1859, comparative morphology was already a highly developed discipline, concerned with the relationship of form to function, with problems of classification, with the interpretation of fossils, with embryology or development, and so on (Russell, 1982). Darwin gave a new impetus to morphology by arguing that all those topics get their interest or significance from history, through descent of species by modification from common ancestors. Seen in this light, comparative study of structure, of embryology and of fossils became the key to reconstructing phylogeny, the course of evolutionary history, and biologists eagerly took up that work in the late nineteenth century. But by the early twentieth century, phylogenetic reconstruction began to fall into disrepute, partly because its results were seen as too speculative, and partly because the main outline of phylogeny was thought to have been worked out, with only trivial details or minor branches remaining to be filled in. Interest shifted from phylogeny, the pattern of genealogy, to more lively fields like genetics and the mechanism of evolution, or ecology and the process of community development. By about 1960, it seemed that few beyond the palaeontologists had any real interest in phylogenetic reconstruction: Medawar's (1967, pp. 22, 23) comments on the status of the work give a flavour of the times: 'nearly all the great dynasties in the history of animals were known', all that remained was 'research in the parish registers of evolution'.

For, by then, molecular biology had arrived, to revolutionize almost the whole science of life. With understanding of the role of DNA in heredity and in manufacture of proteins came the realization that every organism carries a record of its history encoded in the nucleotide helices of its DNA and the amino-acid chains of its proteins. It seemed, too, that this molecular record of phylogeny must be less cryptic than the record hidden in morphology and in the fossil record. At first, the hard part was to get at this molecular record. The simplest and most venerable technique for extracting comparative information from molecules is immunology, a method of comparing serum proteins by their reaction to an antiserum. Though relatively crude, immunological comparisons gave valuable pointers to phylogenetic relationships long before anyone knew why. We now know that the strength of the immunological reaction between two proteins is an approximate measure of the match between the amino-acid chains that give their primary structure. This became clear in the 1960s, first when the technique of electrophoresis was introduced to allow rapid and direct comparisons between proteins from different species, and later when the first few amino-acid sequences of proteins were laboriously worked out. For example, immunological comparisons showed that the antigenic distance between human and chimpanzee serum proteins is very small, and this matched the results from protein sequences as they became available during the 1960s and 1970s: humans and chimpanzees have identical sequences in cytochrome *c* and in alpha and beta haemoglobin, while in myoglobin and delta haemoglobin there is only a single difference in each chain. Given comparable sequence information from other species, there is a wealth of information potentially bearing on phylogeny. Two of the contributors to this book, Walter Fitch and Morris Goodman, were among the pioneers in developing methods of building phylogenies from protein sequence data. Molecular parish registers gave phylogeny a new respectability.

By the late 1970s, techniques of protein sequencing were becoming fairly rapid and semi-automated. But at about that time, protein sequences began to seem old-fashioned with the development of methods for rapid sequencing of DNA itself. These techniques depend on cloning – inserting the portion of DNA to be sequenced into a bacterial virus, which can then be cultured to yield innumerable copies to work with – and on techniques blocking synthesis of these copies at particular points, giving fragments of different lengths which can be separated by electrophoresis and so read off. Once the DNA has been cloned, it is now possible for one worker to sequence something like 5000 nucleotides ('letters' in the DNA code) in a week or so. Of course, DNA sequences are now coming in a torrent, and in the next few years we can expect them to provide an almost overwhelming quantity of data bearing on phylogeny and other aspects of evolutionary history. Before we are overwhelmed, this book is an attempt to take stock.

For molecular phylogenetics by no means has things all its own way; DNA or protein sequencing, DNA hybridization and other modern molecular techniques have not yet pushed comparative morphology into the shadows. That is obvious from virtually every chapter in this book. One reason why morphologists are willing to stand up for their discipline is that the growth of molecular phylogenetics has been accompanied (coincidentally, I believe) by a second flowering of interest in morphological phylogenetics. To understand why that should be, and to provide a background to some of the arguments presented in this book, it is best to go back to the beginning, and to see how the methods of phylogenetics have developed.

Homology and analogy

The basic conceptual tools of comparative morphology are homology and analogy, the names Richard Owen gave (in 1843) to two relations that must have been recognized for as long as morphology has been studied. Distinguishing homology from analogy can be child's play: the human arm, the front leg of a horse and the wing of a bird are all homologous; the wing of a bird and of a butterfly are analogous. Here the three homologues exemplify 'the same organ ... under every variety of form and function' (Owen's 1843, p. 379, original definition of homologue), and the two analogues exemplify 'a part or organ ... which has the same function' (Owen's first definition of analogue). What about the wing of a bird and the wing of a bat? Experienced biologists may give different answers when asked to name the relation between those structures; the wings of birds and bats are homologous as forelimbs, and analogous as wings.

For Richard Owen and other pre-Darwinian morphologists, discriminating homology – 'true' or 'essential' correspondence – from analogy – superficial or misleading similarity – was the key to discovering the natural hierarchy. Their goal was to make the diversity of life comprehensible as variation on common themes or plans, summarized in the groupings of a classification. Owen called the common plan of each group its archetype, a sort of blueprint comprising the homologous features shared by all the group's members. In *The Origin of Species* (1859) Darwin pointed out that the natural hierarchy discovered by Owen and other systematists, 'the grand fact in natural history of the subordination of group under group', was fully explained by his theory (Darwin, 1859, pp. 413, 420):

> on the view that the natural system is founded on descent with modification ... the characters which naturalists consider as showing true affinity between any two or more species [i.e. homologies], are those which have been inherited from a common parent, and, in so far, all true classification is genealogical; that community of descent is the hidden bond which naturalists have been seeking, and not ... the mere putting together and separating objects more or less alike.

Clearly presented here is a programme for research: evolution makes sense of the distinction between homology and analogy by seeing homology as evidence of common ancestry, and analogy as evidence of convergent evolution towards more or less similar form from dissimilar ancestry. Thus, the example of birds' and bats' wings is explained by the descent of birds and bats from a common ancestor furnished with forelimbs (a tetrapod), but not with those limbs modified as wings.

So, in the decades after *The Origin of Species*, comparative morphology became phylogenetic research; common ancestors replaced archetypes, and homology became evidence of common ancestry rather than of common plan. But these were changes in doctrine rather than in practice: the conclusion about forelimb structure in the common ancestor of birds and bats was reached not by observing that ancestor, but by reconstructing it as an abstraction in the same way that Owen reconstructed his archetypes, from the homologous features characterising systematic groups. And Darwin's expectation in *The Origin* that 'Our classifications will come to be, as far as they can be so made, genealogies' (1859, p. 486) did not seem to be met. Phylogenetic research somehow failed to produce the goods, and so fell into disrepute. In Ernst Mayr's great book on the history of biology (1982, p. 218), he says of 'our understanding of the relationships of the higher animal taxa' that

> Honesty compels us to admit that our ignorance concerning those relationships is still great, not to say overwhelming. This is a depressing state of affairs considering that more than one hundred years have passed since the great post-*Origin* period of phylogenetic reconstruction. The morphological and embryological clues are simply not sufficient for the task.

That is one way of looking at the problem: it is a practical one, and the clues are poor, the trail is cold. Another way of looking at the problem is as one of theory: can our distinction of homology from non-homology (analogy or homoplasy) be improved? Biologists are still intrigued by that question, for papers on homology are still commonplace (a few interesting recent ones: Ghiselin, 1976; Rieppel, 1980; Patterson, 1982; Van Valen, 1982; Roth, 1984), and the subject has cropped up in the correspondence columns of *Nature* during the last few years.

Phenetics and cladistics

Whatever the case, progress with phylogeny was slow to come, and dissatisfaction was general. No doubt stemming partly from that dissatisfaction, there appeared in the 1960s two new and very different approaches, now called phenetics and cladistics. Phenetics, or *numerical taxonomy* as the school was originally called (Sokal & Sneath, 1963; also Sneath & Sokal, 1973), acknowledged that phylogeny was largely unknown and perhaps unknowable: classification by genealogy might be a laudable goal, but not one that was attainable. The pheneticists therefore proposed objective and repeatable procedures by which similarities between organisms or 'operational taxonomic units' (OTUs) might be evaluated. The main tenets of phenetics were that as many 'characters' as possible should be compared (preferably 100 or more), that all characters should be given equal weight, and that relationships between OTUs should be expressed as an index of similarity, a measure of phenetic distance derived by an algorithm from the matrix of characters. It must be said on behalf of phenetics that overall similarity is probably usually a reliable guide. Most of the groups that we accept as real (butterflies, insects, birds, mammals, etc.) are recognizable on grounds of similarity. And 'phenetic' groups that are unreal, such as 'fishes' including whales (mammals) and lungfishes (related to tetrapods), evaporate if enough characters are evaluated. But phenetics retreats from theory, from grappling with how to discriminate homology and analogy; and its hopes for operationalism – for safe, sure knowledge – are surely mistaken.

Whereas phenetics stems from belief that phylogeny is unknowable and genealogical classification therefore a pipedream, cladistics developed from a conviction that since genealogical classification is imperative, procedures must be found that will permit it. The principles of cladistics, originally called *phylogenetic systematics*, were developed by the German entomologist Willi Hennig (1913–76). He first published his ideas in a 1950 book, but they did not catch on until an English-language (and much revised) version was published (Hennig, 1966). Among Hennig's contributions, I single out four as particularly influential. Because Hennig's concepts and terms are used by most contributors to this book, it is worth explaining at least these four ideas (Fig. 1.1).

1. A clear definition of relationship, in terms of recency of common ancestry. From this definition flow other concepts, in particular those of monophyletic and non-monophyletic groups (Fig. 1.1).
2. A clearly argued method of detecting relationship, by means of

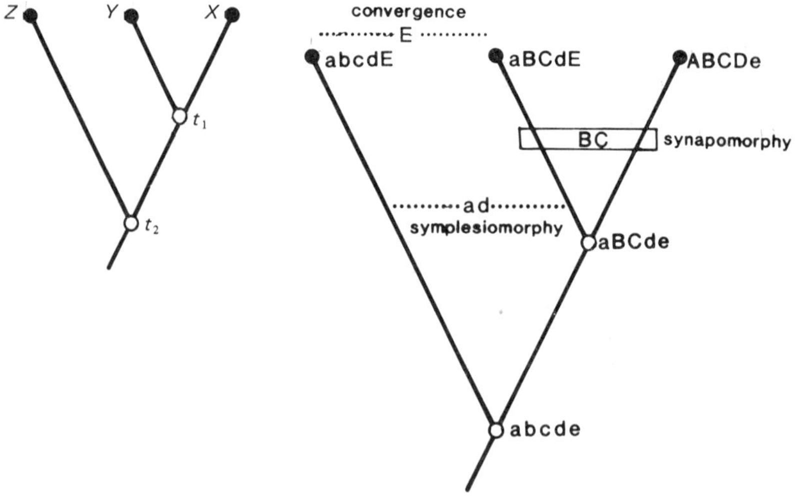

Fig. 1.1 Hennig's concepts of relationship, of how it is detected, and of kinds of group.

Left: Relationship: 'A species X is more closely related to another species Y than it is to a third species Z if, and only if, it has at least one stem species in common with species Y [at t_1] that is not also a stem species of Z' (Hennig, 1966, p. 74). The sole criterion is recency of common ancestry.

Right: Detection of relationships by distribution of characters in species Z, Y and X. Characters observed in the three species and inferred in their common ancestors (open circles) are listed beside each, with lower case letters implying primitive states and capitals derived states (i.e. 'A' is the derived or advanced homologue of 'a'). *Synapomorphies* are shared derived characters, homologous features characterizing monophyletic groups. Shared primitive characters, *symplesiomorphies*, are homologous features inherited from a more remote stem species, so that their distribution is wider than the group of immediate interest. *Autapomorphies* are a third class of character, those unique to a species or group, e.g. 'A' and 'D' in species X.

Monophyletic groups, those characterized by synapomorphies, can also be defined in terms of relationships: a monophyletic group (e.g. $X + Y$) contains species which are more closely related to each other than to any species classified outside the group. Non-monophyletic groups, which fail that definition (e.g. $Z + Y$), may be of two sorts: if they are defined by shared primitive homologies, symplesiomorphies like characters 'a' and 'd' in the diagram, they are *paraphyletic* (e.g. invertebrates, fishes, reptiles); if they are defined by convergent characters, like 'E' in the diagram, they are *polyphyletic* (e.g. a group including fishes + whales and dolphins, or birds + bats).

shared derived characters (which Hennig called *synapomorphies*), homologies characterizing monophyletic groups, or features assumed to have originated in the stem species of a group.

3 A simple method of expressing relationships and character distribution, in dichotomous branching diagrams (Fig. 1.1), now called cladograms. When entered on a cladogram, Hennig's three terms symplesiomorphy, synapomorphy and autapomorphy are seen as successively more restrictive classes of homologies, so that characters change their status according to the generality of the problem in hand. To the anthropologist, reduction of the tail to a vestige during early ontogeny is symplesiomorphous in humans; to the student of primates, reduction of the tail is a synapomorphy of hominoids (apes and humans); to the mammalogist, reduction of the tail is autapomorphous in various mammalian subgroups such as hominoids and sloths; and to all of these, *development* of a tail is symplesiomorphous, characterizing chordates.

4 A research programme for phylogenetics is implicit in the use of dichotomous branching diagrams and in Hennig's term 'sister-group'. We expect that every species or monophyletic group of species has a sister-group, i.e. one species or group which is its closest relative. The job of working out the phylogeny of any species or group is best approached by searching for its sister-group. Although Hennig did not emphasize it, the difficulties of that job are underlined by the number of different dichotomous cladograms possible for even a small number of species or groups (terminal taxa): for 3, 4 and 5 terminal taxa there are respectively 3, 15 and 105 cladograms; for 9 taxa there are over 2 million. Cladograms are not necessarily equivalent to evolutionary trees; for instance, trees may contain named ancestral species. Felsenstein (1978) calculated that there are 262 possible trees for 4 terminal taxa, and those 262 trees were analysed in detail by Nelson & Platnick (1981). It is useful to think of cladograms as sets of trees: for example, each of the 15 possible dichotomous 4-taxon cladograms can be thought of as a summary of a set of about a dozen trees (Fig. 1.2). All these numbers apply to *rooted* trees; unrooted trees or networks, like Fig. 7.1, are a bit less prolific, with respectively 1, 3 and 15 unrooted trees for 3, 4 and 5 terminal taxa.

These and other developments in cladistics are expounded in texts such as Nelson & Platnick (1981), Wiley (1981) and Ax (1987), and need not be taken further here. Since Hennig's ideas began to exert an influence, in the late 1960s, cladistics has been the source of much controversy, and of some unusually acrimonious debate among biologists. Why that should be will make an interesting chapter in the sociology of science. For the moment, I note one further aspect of Hennig's methods which explains at least part of the acrimony, and is relevant to the development of phylogenetics. Though he defined relationship in terms of stem species (Fig. 1.1), and tied his concepts of monophyly and non-monophyly directly to common ancestry, Hennig saw common ancestors, the nodes in his cladograms, as theoretical necessities, hypothetical species whose properties one might infer; he did not see them as practical problems, species that one might try to identify or have to classify. (This is a version of the distinction between cladograms and trees, Fig. 1.2.) Cladists treat species, whether living or fossil, as the tips of branches in cladograms, not as potential ancestors one of another. This style of analysis was seen by some palaeontologists as a threat or insult to their subject. (By about 1960 palaeontology had achieved such a hold on phylogeny reconstruction that there was a commonplace belief that if a group had no fossil record its phylogeny was totally unknown and unknowable.) On the contrary,

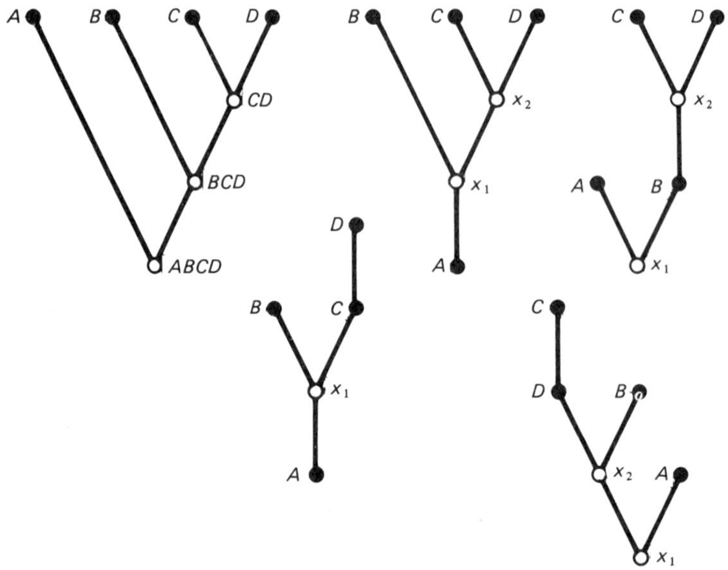

Fig. 1.2 Cladograms and evolutionary trees. At top left is one of the fifteen possible dichotomous cladograms for four species (A, B, C, D). The three nodes linking the four species can be thought of as symbolizing the synapomorphies of the three nested monophyletic groups $\{A\{B\{CD\}\}\}$. The other diagrams are examples from the set of 12 evolutionary trees compatible with this cladogram. They differ from it by including identified or unidentified (X_1, X_2) ancestral species.

Hennig's methods are a procedure for reconstructing the history of groups of any kind, whether they include no fossils, or nothing but fossils, or a mixture of fossil and living species. This view – non-terminal branches of the evolutionary tree as hypothesis – was one cause of the dispute over cladistics; much more important, this view made the methods, results and style of argument of morphological phylogeneticists directly comparable with the work of phylogeneticists dealing with molecular data.

Molecular sequences: homology becomes a statistical concept

Protein sequence data began to accumulate during the middle 1960s. Molecular biology has developed, unlike comparative morphology, entirely within the evolutionary tradition; there is no pre-Darwinian background. In other words, molecular data present something entirely new, undreamed of by Darwin and his predecessors. But though new, molecular data fall comfortably within the tradition of comparative morphology. Given sequences from two (or more) species (or individuals), the first question to ask must be whether they are homologous, so that it is useful or informative to compare them. Or are the sequences merely analogous, so that comparisons will be misleading? In order to follow how such decisions are made with molecular data, a short digression on terms may help. Richard Owen, wrestling with the problem of defining homology, of distinguishing essential from inessential sameness, eventually settled on identity in name: 'A "homologue" is a part or organ in one organism so answering to that in another as to require the same name' (Owen, 1866, p. xii). In other words, homologies are 'anatomical singulars' (Riedl, 1978, p. 52), structures of which there is only one, or a bilateral pair, per organism (instances in mammals are femur, pancreas, lower canine tooth). 'Anatomical singulars' exemplify the morphologist's commonest usage of homology, and this singularity permits one decisive test of hypotheses of homology. If a structure in one organism or group is thought to be homologous with some structure in another, that hypothesis will be disproved if the two are found together in one organism. For example, our belief that the human arm (a mammalian forelimb) and the wings of birds are homologous would be shown to be false if angels (with both arms *and* wings) were discovered.

There is another kind of morphological homology, between anatomical plurals rather than singulars, when several or many copies of a structure exist in each organism (examples are red blood cells in vertebrates, and leaves in flowering plants). Multiple homologues of this type are called *homonyms* when it is necessary to distinguish them from the singular kind of homologue. At the molecular level, virtually all homologies are technically homonyms, since the homologous molecules exist in multiple copies per organism (the only exceptions here might be parts of the genome of prokaryotes and unicellular eukaryotes). So most molecular homologies are homonyms.

Molecular biologists inherited the term homology from classical evolutionary morphology, and in the early days of molecular phylogenetics one can find foreshortened versions of debates which occupied decades in classical biology: Can homology be made operational? Is homology more than similarity? Is the criterion of common ancestry illusory because it leads to a circular argument? Walter Fitch neatly summarized one proponent's position in these debates by adapting a quotation from *Alice's Adventures in Wonderland*:

> 'Homology? Homology?' and sometimes 'Analogy?' for you see, as she couldn't know the answer, it didn't much matter which word she used for it. (Fitch, 1970, p. 112)

The main difference between this debate among biochemists and those among morphologists is that the concept of homology between molecular sequences became quantifiable. A DNA sequence can be thought of as a number to base 4 (since there are 4 possible nucleotides at any position in the sequence); in the same way, a protein sequence can be thought of as a number to base 20 (since there are 20 possible amino acids at any position). If two DNA sequences are aligned, or written out side by side, they will show matches at about 25 per cent ($\frac{1}{4}$) of positions by chance alone (assuming that all four nucleotides occur with equal frequency). But the probability that two sequences of only five nucleotides will match is about $\frac{1}{1000}$, and the probability that ten will is less than one in a million. Two amino-acid sequences should match by chance at about 5 per cent ($\frac{1}{20}$) of positions (if all 20 amino acids occur with equal frequency). When matches occur more often than is expected by chance, an explanation is needed, and the two available explanations are the classical ones of homology (resemblance caused by divergent descent from the same common ancestor) and analogy (resemblance caused by convergent descent from different ancestors).

But arriving at the data to be explained, or deciding if anything needs explaining, is not merely a matter of aligning sequences and counting matches between them. The match between two sequences can be improved if 'gaps' are allowed, where a gap implies insertion of DNA in one lineage, or deletion in the other. If gaps are permitted *ad lib.*, amino acid sequences can be made to match at around $\frac{1}{4}$ of positions (Barker & Dayhoff, 1972, p. 101), and if the two sequences are sufficiently long, they can be rendered almost identical (Doolittle, 1981). So it is necessary to 'penalize' gaps, by weighting them more heavily than ordinary point mutations, which change one nucleotide in a DNA sequence. Further, in known sequences the 4 nucleotides and 20 amino acids do not occur with equal frequency, and structural features of unrelated proteins may increase matching between them. Nevertheless, these and other complications can be allowed for, and searching for homology among nucleotide or amino acid sequences has become a statistical problem, tackled by computer programming: one criterion for sequence homology is matching 3.0 or more standard deviations above chance expectation, but more sensitive measures exist for proteins (Doolittle, 1981) and DNA sequences (Karlin *et al.*, 1983).

Alignments with less in common might still be homologous, and when comparisons of primary (sequence) structure of proteins fail to reveal relationship, resemblance in tertiary (three-dimensional) structure is often taken as evidence of homology. Phillips, Sternberg & Sutton (1983, pp. 168–9) write 'during evolutionary change, three-dimensional structure is conserved more strongly than amino-acid sequence' so that 'the only surviving evidence for divergent evolution sometimes may lie in the three-dimensional structure of the protein'. But here we are back to raw similarity as the criterion of homology, and these statements carry the implication that evidence of common ancestry may persist in phenotypes (proteins) when it has disappeared in the hereditary material (DNA sequences), as if homology were manufactured afresh in each generation. 'Spatial equivalence' seems a better term than homology for resemblance in three-dimensional structure of proteins (W.M. Fitch, personal communication).

In general, given two or more homologous sequences, the problem of reconstructing phylogeny becomes one of reconstructing ancestral sequences which will produce the descendent (i.e. given or sampled) sequences with the *minimum* of evolutionary change. Minimizing inferred change is not meant to say anything about the evolutionary process, or that it takes the shortest route. Minimizing change is merely the principle of parsimony, i.e. that observations be explained with as few assumptions as possible. Parsimony – Occam's razor – is a principle of reasoning, not an inference about nature (though it is sometimes used in the latter sense, and it is always worth thinking out the level at which parsimony is being applied). There are also those, including Bishop & Friday (Chapter 5), who argue that we cannot reconstruct phylogeny without a model of the evolutionary process. Their probabilistic model and maximum likelihood method may give different solutions to those produced by parsimony analyses.

Whichever method is used, the quantity of characters provided by sequence data combined with the quantity of possible cladograms (for more than a handful of species) means that the task of finding the most parsimonious or best-supported cladogram is left to the computer. For even a few species, this may stretch or exceed the capacity of available computers: McKenna (Chapter 3) notes that one phylogenetic analysis of 5 proteins in 11 species occupied 150 hours of computer time, while Bishop & Friday's method (Chapter 5) needed

40 minutes for 1 protein in 6 species, and examining all possible cladograms is 'computationally impossible' beyond about 10 species. McKenna's analyses of sequence data were done by hand, not computer, and cover more than 50 species. His comments on the merits of manual methods are interesting, as is the fact that no computer is (as yet) capable of checking that his solutions are the best available.

Orthology and paralogy

One of the most important conceptual developments in molecular phylogenetics is Walter Fitch's (1970) distinction between two kinds of homology in proteins or in genes, *orthology* and *paralogy*. Orthology has the same meaning as traditional morphological homologies, like the forelimb or the femur; orthologous genes and gene products are found in different species, and are the basis for inferences of common ancestry among species. Paralogy has more the significance of two *different* morphological homologies, like the forelimb *and* the femur; paralogous genes and gene products may be found in the same organism, and are the basis for inferences of common ancestry among genes, not species. In some ways, paralogy is like serial homology, the name that Richard Owen introduced for the relation between repeated structures in an animal, such as the limbs of a crustacean or the vertebrae of a mammal. But paralogy differs from serial homology in that it is explained not by duplication in ontogeny (as the rudiments of a crustacean's appendages or a mammal's vertebrae develop), but by duplication of genes in phylogeny. For example, the globin family of genes in vertebrates contains genes coding for myoglobin, alpha, beta and gamma haemoglobin, several other embryonic or foetal haemoglobins, and other genes (Jeffreys *et al.*, 1983). A phylogeny of vertebrates could be constructed from the amino-acid sequences of myoglobins from some species, alpha haemoglobins from others, betas from others and gammas from still others, and the phylogeny would be radically different from that built up by morphologists. A different pattern would be evident in it, a pattern in which the alpha haemoglobins were grouped, the myoglobins were grouped, and so on. Such a phylogeny would confuse the pattern of descent of *genes* (paralogy) with that of *species* (orthology). The explanation for paralogy is gene duplication (Ohno, 1970), the theory that diversity of genes has arisen within genomes by mutations doubling parts of chromosomes which have then diverged over the course of time, a process analogous to speciation and the divergence of sister-species over time. The importance of the concept of gene duplication in evolutionary theory is that it permits an escape from the iron grip of stabilizing selection. Duplicate copies of a gene are no longer the only source of that gene's product for the organism, and selection may be relaxed, allowing one of the copies to accumulate mutations that would be deleterious in a unique gene.

The distinction between orthology and paralogy brings with it one problem for molecular phylogenetics: given a set of homologous sequences from different species, one can construct a phylogeny by parsimony or likelihood methods, but how can one discriminate a phylogeny reflecting divergent descent of species (orthologous sequences) from one reflecting divergent descent of genes (paralogous sequences), or from a mixture of the two? As with the angel's wings and arms, the test should be simple: paralogous sequences will coexist in the same organism, orthologous sequences will not. But molecular genetics had more surprises in store.

Introns and exons: partial homology

First came the discovery, in the late 1970s, that there is rarely a direct match between the gene, the sequence of information in DNA, and the protein, the expression of that information by translation into an amino acid chain. In most eukaryote genes there are several translated or expressed portions, the exons, separated by untranslated portions, the introns, which are transcribed into RNA but 'spliced out' before translation into amino acids. Introns are generally much longer than exons, so that genes (introns + exons) are often about 20 times larger than would be predicted from the length of the protein (exons alone). It also turns out that exons, the coding (translated) sequences of the genes, correspond to structural, and sometimes to functional, domains or sectors in the protein. Given the tertiary (three-dimensional) structure of a protein, it is possible to predict the sites of exon/intron junctions (Gõ, 1981). For example, in vertebrate globin genes there are three exons separated by two introns, but structural considerations predict that the long central exon should be divided by an intron; this 'missing' intron was discovered in a plant globin, leghaemoglobin from soybean (Jensen *et al.*, 1981; Phillips *et al.*, 1983).

Given this knowledge about the relation between gene structure and protein structure, one can visualize how new proteins with new functions might arise by exon duplication and by 'exon shuffling': chromosome mutations could produce new conjunctions of exons from different genes, so joining portions of different proteins in novel combinations, and occasionally generating new functions (Gilbert, 1978, 1985; Darnell & Doolittle, 1986). Thus, just as gene duplication can explain paralogy between genes or proteins within one genome, so exon shuffling can explain 'partial homology', a situation in which a gene or protein might combine regions (exons) homologous with two (or more) different genes or proteins. Such hybrid genes must originally arise within one genome and so exhibit partial paralogy, but over time they might be involved in speciation and divergent evolution and so become partially orthologous (Fig. 6.4).

Morphological homology is an all-or-none relation: structures, or behaviours, or whatever are either homologous or non-homologous, and partial homology does not occur. Molecular homology seems to be more subtle, and orthology, reflecting species phylogeny, paralogy, reflecting gene phylogeny, and partial orthology or paralogy must all be discriminated.

Pseudogenes – hidden paralogy; foreign genes – xenology

Pseudogenes, first discovered in 1980, give the story yet another twist. A pseudogene is a segment of DNA which is paralogous with a functional gene, but is not transcribed or not translated because of some defect: if gene duplication is analogous to speciation, pseudogenes are 'dead genes', analogous to extinct species. Many pseudogenes seem to have been produced by reverse transcription from messenger RNA; they are silent or extinct because they lack the promoter regions which initiate transcription in DNA. Other pseudogenes are normal genes, extinguished by one or more mutations which have made them biologically inactive or silent, for example by introducing a stop codon, or shifting the reading frame, or altering an RNA processing signal. In order to survive in a lineage, these pseudogenes must represent originally redundant duplicate copies of genes; otherwise, the mutations that have silenced them would be deleterious and so quickly eliminated.

From the viewpoint of homology and phylogeny reconstruction, pseudogenes might present one more problem. They seem to be abundant: Hunkapiller *et al.* (1982) estimated that 25 per cent of globin genes are pseudogenes, and this proportion is likely to increase rather than decrease as work progresses. Given the abundance of pseudogenes, how can one infer that functional genes in different species – for example, the single beta haemoglobin gene of humans and other higher primates (Jeffreys *et al.*, 1983) – are orthologous? The possibility must always exist that a given pair is paralogous, one member of each of two truly orthologous pairs having been silenced by conversion into a pseudogene. As a result, one might compare two sequences believing that they had diverged since the split between two lineages, whereas they had actually diverged from a gene duplication before that split.

Whereas pseudogenes raise the possibility of confusing paralogy with orthology and *under*estimating the age of divergence between two sequences, there is another newly found homologous relation between sequences which is neither orthology nor paralogy, and might cause *over*estimation of times of divergence. The process involved here is transfection, 'horizontal' transfer of heritable material between lineages. This sort of transfer is best understood in prokaryotes, where the agents responsible for transfection are plasmids or bacteriophages. In eukaryotes, similar events probably occur, but are much harder to study. Horizontal transfer of genes or fragments of DNA from one lineage to another means that one might compare sequences from different species that are neither orthologous (reflecting phylogeny of species) nor paralogous (reflecting phylogeny of genes within those species) but are still homologous (reflecting common ancestry). Gray & Fitch (1983) proposed the name *xenology* for the relation involving 'foreign genes' acquired by transfection, horizontal transmission between lineages which is necessarily more recent than the vertical transmission by descent of paralogous or orthologous genes.

In practice, mistaken inferences due to silenced genes or foreign genes may not be common problems in inferring phylogeny. But abundant pseudogenes certainly exist, and pseudogenes may themselves be duplicated or amplified *after* silencing (Ghazal, Clark & Bishop, 1985). These silenced members of families of paralogous genes are one more element in the developing picture of eukaryotic genomes not as monolithic strings of genes but as more fluid things, 'islands of transcribed sequences in a sea of silent DNA' (Loomis & Gilpin, 1986, p. 2147).

Clocks and neutrality

Once a gene has been silenced, what will be its future? Since the pseudogene (as it now is) does not contribute to the phenotype, it should be immune to natural selection, of either the purifying or stabilizing type (which eliminates variants and preserves the *status quo*), or the directional type (which promotes change). Mutations in pseudogenes will therefore be neutral so far as selection is concerned, and may increase in frequency over the generations or be eliminated by chance, by random genetic drift.

The theory which predicts the behaviour of pseudogenes and other components of 'the sea of silent DNA' is the neutral theory of molecular evolution, associated in particular with Motoo Kimura (1979, 1983). The roots of neutral theory extend back into the 1960s, to observations on protein sequences and on genetic variation within species. In the mid-1960s, the introduction of electrophoresis revealed huge and unexpected reservoirs of genetic diversity within populations. Individuals in most species are heterozygous for around 5–15 per cent of their genes. How is this variation maintained? Two explanations are available, natural selection (principally balancing selection favouring increased fitness of heterozygotes) and random drift (where the variants represent neutral mutations maintained by recurrent mutation and chance propagation). In favour of the neutral explanation is the cost or genetic load of natural selection: if balancing selection is maintaining heterozygosity in a population, less fit homozygotes must occur and be selectively eliminated in each generation. Kimura calculated that if 2000 loci are maintained in this way each with only 1 per cent heterozygote advantage (not unreasonable estimates), every individual must produce about 22 000 offspring for population numbers to remain constant! To date, despite abundant data and sophisticated theorizing, the neutralist–selectionist issue has not been resolved where intraspecific polymorphism is concerned: as Nei & Graur (1984, p. 73) put it, the 'null' hypothesis of neutral mutations has not yet been rejected. But in my view, neutral theory is favoured by its success in explaining observations on protein and nucleotide sequences.

In the early days of protein sequences, it was soon evident that globins (for example) from very dissimilar lineages differ by virtually equal amounts. One of Kimura's examples compares the alpha haemoglobin sequence of a shark with those of carp, salamander, chicken, mouse and human. The number of amino acid differences from the shark sequence is: carp 85, salamander 84, chicken 83, mouse 79, human 79. Those numbers are approximately the same, implying that the rate of change of alpha haemoglobin has been the same, on average, in the lineages leading to each of these vertebrates from their common ancestor, which must have existed in Silurian times or earlier, more than 400 MyrBP. The same constancy of molecular evolution emerges from Kimura's comparison of the paralogous human alpha and beta haemoglobin sequences, and of the orthologous alpha haemoglobin sequences of human and carp. The differences between the two human haemoglobins can be accounted for by 147 point mutations in DNA, and those between human and carp alpha chains by 149, virtually the same number. The two paralogous human genes have diverged from a gene duplication in early gnathostomes, a little before the divergence of the orthologous actinopterygian (carp) and sarcopterygian (human) genes, again more than 400 MyrBP. In contrast to that inferred constancy of molecular rates, it is evident that the morphological rate of change has been quite different in the lineages leading (for example) to shark, carp, chicken and human. Modern sharks look much the same as their Devonian fossil relatives, whereas the morphology of birds and mammals is totally transformed when compared with their Devonian relatives, the early lungfishes and other sarcopterygian fishes.

Generalizing these examples, it seems that each protein has a roughly constant rate of change, but that rates vary widely between proteins. If the rate for haemoglobin is taken as unity, that for myoglobin is about 0.75, cytochrome *c* is about 0.25, and fibrinopeptides change much faster, about seven times as fast as haemoglobin. Inferences like these are the basis for the molecular clock, the theory that genomes evolve at a constant, clock-like rate. Since it was first proposed in the mid-1960s, the clock hypothesis has been one of the most controversial topics in evolutionary theory. In this book it finds some enthusiastic supporters: 'the molecular clock is perhaps the most important addition to evolutionary doctrine since the time of Darwin' (Carl Woese, p. 181), and for Sibley & Ahlquist (p. 99) the clock is an empirical fact, modified only by effects of generation time. Others are less enthusiastic (e.g. Andrews, p. 28), and Goodman, Miyamoto & Czelusniak (p. 164) are particularly strongly opposed to clock-like evolution of globins and other proteins.

An explanation for a molecular clock can be found in one of the principles of neutral theory, that functionally less important molecules or parts of a molecule evolve faster (in terms of mutant substitutions) than more important ones (Kimura & Ohta, 1974), or 'the rule is that those molecular changes that are less likely to be subject to natural selection occur more rapidly in evolution' (Kimura, 1983, p. 32). If genomes are truly 'islands of transcribed sequences in a sea of silent DNA', then only the islands (and their immediate surroundings) are likely to be subject to natural selection, and the sea of silent DNA is free to evolve at a rapid, and constant, rate. The cause of this rapid evolution is chance: at each site in DNA mutations will arise and will occasionally spread through the population to fixation, *unless* that site is subject to functional constraint, making some contribution to the fitness of the organism which requires that it remain unchanged. Evidence for this proposition has accumulated during the last few years, as DNA sequences have become available.

First, in comparisons of homologous DNA sequences from coding regions, it is always found that the DNA divergence is greater (usually much greater) than the amino-acid divergence. The excess divergence in the nucleotide sequences is due to synonymous or *silent* substitutions, point mutations which have no effect on the amino acid specified because of the degeneracy of the genetic code. For example, valine is specified by four codons, GUA, GUG, GUU, GUC, and any change in the third position will be silent, whereas any change in the first two positions will be non-silent, altering the amino acid. Arginine, serine and leucine, each with six codons, can sustain one silent change in the first position, as well as third position changes. Methionine and tryptophan, each with only one codon, cannot show silent substitutions. In general, many third position changes are silent, as are a few in the first position, and second position changes are non-silent (cf. Chapter 3, Appendix 2). The inferred silent rate of evolution is always higher, usually by a factor of between 2 and 4, than the non-silent rate. This can be explained by selective constraint on non-silent substitutions, whereas silent substitutions accumulate at close to the neutral rate. That conclusion is consistent with the fact that the silent rate of evolution seems to be about equal in different genes, whereas the non-silent rate (the rate of change in amino acids) varies widely from one protein to another.

Second, in comparisons of homologous DNA sequences comprising both coding regions (exons) and intervening sequences (introns), it is always found that introns show greater divergence than exons. Again, this can be explained by the weaker selective constraint on introns (spliced out before translation) than on coding regions.

Thirdly, in comparisons of orthologous sequences of genes and of pseudogenes, it is always found that pseudogenes show far more divergence than their functional paralogues, implying more rapid evolution. And whereas substitutions in functional genes are dominantly silent, and therefore concentrated in the third position of codons, substitutions in pseudogenes show no such restriction (Li, Gojobori & Nei, 1981). Pseudogenes seem to show the highest rates of evolution yet estimated in nuclear DNA. So, to return to the question posed at the beginning of this section, what will happen to a pseudogene? Over the generations it will accumulate nucleotide substitutions, and also deletions and insertions, at a comparatively rapid rate, so that its homology with functional paralogues will decay, and should it survive long enough, it will eventually melt into the sea of silent DNA, becoming indistinguishable from a random DNA sequence. During this process of change, it will contribute to clock-like change in the genome.

Molecules versus morphology

With that metaphor – dead genes drifting into a sea of random DNA by a clock-like process of decay, by rapid evolution occurring *in spite of* rather than *because of* natural selection – we are a long way from the obvious and exquisite fit of form to function seen by the morphologist comparing the human arm and the bat's wing. In these terms, perhaps the molecular end of the spectrum seems far more austere or sophisticated than the morphological end. Yet there is a common thread joining gross anatomy to recognition of pseudogenes: all useful comparisons in biology depend on the relation of homology. And in sketching some of the ramifications of molecular homology, I intended to float the idea that molecular homologies are no more secure, and are possibly more precarious, than morphological ones. For instance, in comparing the skulls or the limbs of vertebrates, and most of the parts of those structures, from sharks to mammals, we can be sure that we are dealing with true homologues (that the skull *is* the skull, not sometimes a duplicated or partial version), and therefore that the hierarchies we detect or the phylogenies that we infer in morphology reflect reality. Can we say the same of (say) the alpha haemoglobin molecule from sharks to mammals? Even in thoroughly studied genomes like the human, new members of the alpha gene family are to be discovered (Marks, Shaw & Shen, 1986), and beyond the complications of discriminating orthology (divergence in species phylogeny) from paralogy (divergence in gene phylogeny) there is the possibility of xenology due to gene transfer, and there may be mixtures of orthology and paralogy within one molecule.

That last possibility is taken further by Goodman, Miyamoto & Czelusniak (p. 149) and illustrated in Fig. 6.4: gene conversion or gene correction (where part of one gene is used as a template for a neighbouring gene) could cause mixed orthology and paralogy, or paralogy to masquerade as orthology. No comparable intricacies seem to impede morphological comparisons. Reviewing these difficulties, Goodman *et al.* write (p. 149) 'molecular biologists, in their efforts to reconstruct phylogeny from sequence data, encounter problems comparable to those faced by morphologists'. Bishop & Friday conclude their Chapter 5 with virtually the same message: 'Our analysis seems to emphasize the similar problems encountered in the interpretation of biochemical and morphological data ... both types of data need to be approached with equal caution.' Goodman *et al.* and Bishop & Friday are principally concerned with coding sequences, as evinced in proteins. Andrews (Chapter 2) and McKenna (Chapter 3), both primarily morphologists, come to much the same conclusion as their molecular colleagues: Andrews finds that morphology and molecules are two ends of a continuum, the first richer in functional information, the second in genetic; McKenna writes, 'It should also be obvious that molecular studies can suffer from exactly the same ills that beset comparative anatomical ones: a touchstone has not been found ...'.

So far, virtual unanimity. But Sibley & Ahlquist (Chapter 4) are the odd-men-out among those contributors reviewing vertebrate groups. For them, convergence (analogy, homoplasy) is the besetting difficulty for the morphologist, but with DNA hybridization they have a method that is 'immune to convergence', because only homologous sequences are similar enough to reassociate under their experimental conditions. The result of DNA hybridization experiments is a distance statistic, a number expressing a measure of the difference between the single-copy DNA of two genomes. Other contributors are critical of distance statistics: for Andrews (Chapter 2) they are 'strictly phenetic', for McKenna (Chapter 3) they are 'a dubious mixture of phenetic and cladistic'. Behind those opinions lies a long-running controversy, whose roots in disputes between pheneticists and cladists will be evident from the comments.

Woese's contribution (Chapter 7) deals with prokaryotes, organisms as remote from vertebrates as is possible, and so simple that comparative morphology led nowhere; as Woese puts it, 'microbiology developed essentially without an evolutionary framework' until molecular sequences became available. Woese uses an approach to phylogeny roughly intermediate between distance data (Sibley & Ahlquist's DNA hybridization, for example) and characters (morphological homologies or orthologous sequences, for example). His characters are oligonucleotides in ribosomal RNA, short fragments of about 10 nucleotides which seem to be highly conserved and so recognizable in a wide range of prokaryotes, spanning lineages which must have diverged long before the vertebrates came into existence. A distance measure is derived from comparisons of oligonucleotides, and the distances serve as the basis for phylogenetic trees. By these techniques, Woese and his colleagues have revolutionized our understanding of prokaryote phylogeny: before their work, there was nothing, and now there is a scheme rich in new proposals and in suggestions for new approaches. It may be the same with Sibley & Ahlquist's DNA hybridization phylogenies of birds. Paradoxically, before they started work, ideas of phylogeny among birds, superficially such a popular and well-known group, seem to have been hardly in better shape than those of phylogeny among prokaryotes. As with Woese's work, Sibley & Ahlquist have proposed a wealth of new and unexpected groupings, offering varied opportunities for testing by morphological and other techniques. These proposals will stand or fall by their success in those tests, and by the use that can be made of them as stimuli for further work, not by criticisms of the techniques employed.

Distance data come into focus again with Fitch & Atchley's chapter (Chapter 8). Since they are the only contributors starting out with a known phylogeny (the genealogies of inbred strains of laboratory mice), they are also the only ones able directly to test the performance of morphology versus molecules. Their result is unequivocal: from their molecular data all five of the methods they tried retrieved the correct phylogeny, whereas their morphological data yielded a tree nowhere near the truth. Fitch & Atchley correctly and cautiously temper their conclusion by many caveats. It is to the point here that they are using distance data: at the molecular level, their distance data worked.

Reconstruction of phylogeny, whether from morphological or molecular data, concerns principally the pattern of evolution, whereas the hot topics in evolutionary theory generally concern process, mechanisms that might cause or explain the pattern of phylogeny. In general, the contributions in this book do not have a lot to say about evolutionary mechanisms. But among those specializing at the molecular end of the spectrum, Goodman *et al.* (Chapter 6) insist on the importance of Darwinian evolution, discerning a pattern of rapid adaptive change guided by directional selection succeeded by slowdowns constrained by stabilising selection. Woese (Chapter 7) sees comparable variations in rates of evolution, but leans more towards neutral or non-Darwinian change, with selection as constraint, and he advances a devastatingly simple theory of a link between mutation rate and the tempo and mode of evolution in prokaryotes. Finding ways to test his suggestion that the same mechanism may explain macroevolution in metazoans is just one of the pointers to the future to be found in this book.

REFERENCES

Ax, P. (1987). *The Phylogenetic System.* Chichester: John Wiley & Sons.

Barker, W. C. & Dayhoff, M. O. (1972). Detecting distant relationships: computer methods and results. In *Atlas of Protein Sequence and Structure, 1972, vol. 5*, ed. M. O. Dayhoff, pp. 101–10. Silver Springs: National Biomedical Research Foundation.

Darnell, J. E. & Doolittle, W. F. (1986). Speculations on the early course of evolution. *Proceedings of the National Academy of Sciences USA*, **83,** 1271–5.

Darwin, C. R. (1859). *On the Origin of Species by Means of Natural Selection.* London: John Murray.

Doolittle, R. F. (1981). Similar amino acid sequences: chance or common ancestry? *Science*, **214,** 149–59.

Felsenstein, J. (1978). The number of evolutionary trees. *Systematic Zoology*, **27,** 27–33.

Fitch, W. M. (1970). Distinguishing homologous from analogous proteins. *Systematic Zoology*, **19,** 99–113.

Ghazal, P., Clark, A. J. & Bishop, J. O. (1985). Evolutionary amplification of a pseudogene. *Proceedings of the National Academy of Sciences USA*, **82,** 4182–5.

Ghiselin, M. T. (1976). The nomenclature of correspondence: a new look at 'homology' and 'analogy'. In *Evolution, Brain, and Behavior: Persistent Problems*, ed. R. B. Masterton, W. Hodos & H. Jerison, pp. 129–42. Hillsdale, N. J.: Lawrence Erlbaum Associates.

Gilbert, W. (1978). Why genes in pieces? *Nature, London*, **271,** 501.

Gilbert, W. (1985). Genes-in-pieces revisited. *Science*, **228,** 823–4.

Gō, M. (1981). Correlation of DNA exonic regions with protein structure units in haemoglobin. *Nature, London*, **291,** 90–2.

Gray, G. S. & Fitch, W. M. (1983). Evolution of antibiotic resistance genes: the DNA sequence of a kanamycin resistance gene from *Staphylococcus aureus. Molecular Biology and Evolution*, **1,** 57–66.

Hennig, W. (1966). *Phylogenetic Systematics.* Urbana: University of Illinois Press. (Second edition, 1979.)

Hunkapiller, T., Huang, H., Hood, L. & Campbell, J. H. (1982). The impact of modern genetics on evolutionary theory. In *Perspectives on Evolution*, ed. R. Milkman, pp. 164–89. Sunderland, Massachusetts: Sinauer Associates.

Jeffreys, A. J., Harris, S., Barrie, P. A., Wood, D., Blanchetot, A. & Adams, S. M. (1983). Evolution of gene families: the globin genes. In *Evolution from Molecules to Men*, ed. D. S. Bendall, pp. 175–95. Cambridge University Press.

Jensen, E. Ø., Paludan, K., Hyldig-Nielsen, J. J., Jørgensen, P. & Marcker, K. A.

(1981). The structure of a chromosomal leghaemoglobin gene from soybean. *Nature, London*, **291,** 677–9.

Karlin, S., Ghandour, G., Ost, F., Tavare, S. & Korn, L. J. (1983). New approaches for computer analysis of nucleic acid sequences. *Proceedings of the National Academy of Sciences USA*, **80,** 5660–4.

Kimura, M. (1979). The neutral theory of molecular evolution. *Scientific American*, **241** (5), 94–104.

Kimura, M. (1983). *The Neutral Theory of Molecular Evolution*. Cambridge University Press.

Kimura, M. & Ohta, T. (1974). On some principles governing molecular evolution. *Proceedings of the National Academy of Sciences USA*, **71,** 2848–52.

Li, W.-H., Gojobori, T. & Nei, M. (1981). Pseudogenes as paradigms of neutral evolution. *Nature, London*, **292,** 237–9.

Loomis, W. F. & Gilpin, M. E. (1986). Multigene families and vestigial sequences. *Proceedings of the National Academy of Sciences USA*, **83,** 2143–7.

Marks, J., Shaw, J.-P. & Shen, C.-K. J. (1986). The orangutan adult alpha-globin gene locus: duplicated functional genes and a newly detected member of the primate alpha-globin gene family. *Proceedings of the National Academy of Sciences USA*, **83,** 1413–7.

Mayr, E. (1982). *The Growth of Biological Thought*. Cambridge, Massachusetts: Belknap Press.

Medawar, P. B. (1967). *The Art of the Soluble*. London: Methuen & Co.

Nei, M. & Graur, D. (1984). Extent of protein polymorphism and the neutral mutation theory. *Evolutionary Biology*, **17,** 73–118.

Nelson, G. & Platnick, N. (1981). *Systematics and Biogeography: Cladistics and Vicariance*. New York: Columbia University Press.

Ohno, S. (1970). *Evolution by Gene Duplication*. New York: Springer-Verlag.

Owen, R. (1843). *Lectures on the Comparative Anatomy and Physiology of the Invertebrate Animals*. London: Longman, Brown, Green & Longmans.

Owen, R. (1866). *On the Anatomy of Vertebrates. Vol. 1. Fishes and Reptiles*. London: Longmans, Green & Co.

Patterson, C. (1982). Morphological characters and homology. In *Problems of Phylogenetic Reconstruction*, ed. K. A. Joysey & A. E. Friday, pp. 21–74. London: Academic Press.

Phillips, D. C., Sternberg, M. J. E. & Sutton, B. J. (1983). Intimations of evolution from the three-dimensional structure of proteins. In *Evolution from Molecules to Men*, ed. D. S. Bendall, pp. 145–73. Cambridge University Press.

Riedl, R. (1978). *Order in Living Organisms*. Chichester: John Wiley & Sons.

Rieppel, O. (1980). Homology, a deductive concept? *Zeitschrift für Zoologische Systematik und Evolutionsforschung*, **18,** 315–19.

Roth, V. L. (1984). On homology. *Biological Journal of the Linnean Society*, **22,** 13–29.

Russell, E. S. (1982). *Form and Function* (2nd edn), with a new introduction by G. V. Lauder. University of Chicago Press.

Sneath, P. H. A. & Sokal, R. R. (1973). *Numerical Taxonomy*. San Francisco: W. H. Freeman & Co.

Sokal, R. R. & Sneath, P. H. A. (1963). *Principles of Numerical Taxonomy*. San Francisco: W. H. Freeman & Co.

Van Valen, L. M. (1982). Homology and causes. *Journal of Morphology*, **173,** 305–12.

Wiley, E. O. (1981). *Phylogenetics*. New York: John Wiley & Sons.

2 Aspects of hominoid phylogeny

Peter Andrews

Introduction

The distinction can be made between molecular and morphological approaches to taxonomy both in the type of evidence and the way it is analysed. The molecular approach has the advantage of maximizing genetic information, and for taxonomic relevance it must either cover the whole genome or provide precise sequence information on part of the genome. Either way, the loss of genetic information is small, and it is argued from this that the data have high taxonomic content and that they are best analysed by probabilistic models. Morphology is the phenotypic expression of the genome, and the more complex the structure the less well known is the genetic content. Relatively simple structures like some of the blood group proteins are relatively well understood genetically, whereas the genetic relationship of complex anatomical structures is poorly understood at present. But phenotypic characters can be studied ontogenetically and functionally, and it is possible to study the correlation between such characters in terms of function. It is also common practice to attempt to determine directionality of change so as to distinguish primitive or ancestral characters from specialized or derived characters. Analyses of function, of functional correlation and of polarity can and should be applied to molecular evidence, but at this stage of our knowledge much of our information on function and polarity comes from morphology. It is also unfortunately true at present that the fossil record is exclusively morphological, so that the direct inference of change through time that fossils offer is limited to morphology.

What I propose to do here is to review briefly what I see as a continuum of evidence. I will start at the level of gross morphology, characters which are presumably genetically complex, although we know so little about their genetics that even that is not clear. There is good evidence, however, for the functional correlations between characters. Some knowledge of function is combined with greater genetic information for the blood groups and the chromosomes, and these will be considered next. Finally, the DNA molecule itself has a structure analogous to morphological structures, and the amino acids, and ultimately the proteins, that it codes for can also be viewed in structural terms. The genetic content of the data on DNA and amino acid sequencing is very high, with little loss of genetic information (Frelin & Vuilleumier, 1979) but the functional content is minimal. Having done this, I will then apply the evidence to two events in hominoid evolution, the divergence of man from the last common ancestor with the apes, and the initial divergence of the clades comprising the great apes and man. As far as possible, I will apply all lines of evidence without prejudice and without prior assumptions, although I have to declare an interest which might produce a bias in favour of the morphological end of the spectrum.

The nature of the evidence

The evaluation of character states is the most difficult problem in morphological systematics. Characters shared through common descent are considered homologous, and homologous characters shared exclusively from the last common ancestor of two taxa are considered evidence of relationship between them. This requires the twofold demonstration of uniqueness and homology before any character can be accepted as evidence of relationship, and this requirement, which is difficult enough to establish at the morphological level, where likelihood of convergence in characters can be related to convergence in function, becomes increasingly hard to identify at the molecular level, where convergence in nucleotide substitutions is distinguishable from homology only on the basis of probability.

Morphological character analysis

Anatomical characters are considered both singly and in functional units (Fig. 2.1). A single character of uncertain function and unknown genetics is of limited use in determining relationships. If characters can be demonstrated to be non-functional or conversely can be correlated with some larger functional complex, they would have greater taxonomic value. Single morphological characters are commonly used in two ways: first, as evidence of one particular genealogy, when only those characters supporting one particular view are put forward (Kluge, 1983; Schwartz, 1984); and secondly, when all characters (or as many as possible) relevant to particular branching points are considered, and the choice between different branching orders is made on the basis of shared derived characters (Delson & Andrews, 1975; Harrison, 1982; Andrews, 1985*a*; Martin, 1985*b*).

Characters are rarely, if ever, totally isolated. Correlations between characters may be the result of functional affinity, so that there may be a complex of several apparently separate character states that all relate to the one function

(Andrews & Cronin, 1982). If the correlations are high, they should properly be considered as a single character rather than many, but in this case there are two aspects of the resulting multiple character complex that must be considered. There are, firstly, the function of the character, or what it does, and secondly, how the character complex is constituted. The first relates directly to function, and it is of little taxonomic relevance because convergence in function can occur so readily. Of much greater importance is the way in which the individual characters that constitute the complex combine together to produce the end result. The combination of characters in a complex, and the way they work together, is of greater taxonomic significance, for convergence may be recognized by different combinations of characters producing similar end results. Put very simply, a character complex can be considered convergent in two species if the characters that are part of the complex are different. This is almost a truism, but the converse is important: complexes made up of exactly the same combinations of characters have a high likelihood of being homologous, and the more elaborate the character complex, the more likely is this similarity to be homologous.

Establishing the polarity of change of character states through time is a problem for all characters, and my interpretations are based on outgroup comparisons: I have assumed that gibbons are the outgroup to the great apes

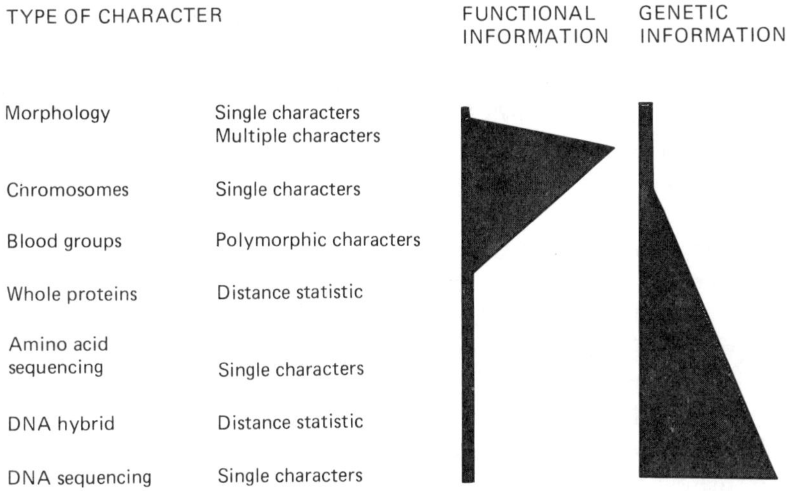

Fig. 2.1 Source of data for reconstructing phylogenetic trees. On the left are shown the data sources discussed in this paper, and next to them is an assessment of the type of information provided by each source, whether single or multiple characters, distance statistics or polymorphisms. On the right are shown graphically the functional and genetic information contents of these data sources (not to scale).

and humans and that the Old World monkeys are the outgroup to the Hominoidea (Andrews, 1985*a*). Two alternatives to outgroup comparisons, complexity and ontogeny, can be mentioned briefly, but they have only limited value in assessing hominoid character polarity. It is probable that the more complex a character is, the less likely is the possibility of reversal. A functional complex, with numerous characters relating to that complex, is likely to be derived with respect to the same complex lacking those characters. I have used this assumption in assessing the relationships of the fossil hominoid *Sivapithecus* with the orang utan (Andrews & Cronin, 1982) and in the present paper in distinguishing between various options in the relationships of the African apes and humans.

The use of ontogeny for determining polarity is problematic in the hominoids. Bonde (1984) has argued strongly in its favour, and he applied it to comparisons between great apes and humans, but it seems to be a considerable over-simplification when applied to human evolution. In Bonde's model, it is more parsimonious to assume that for a pair of species, where one of them changes ontogenetically, the ancestor had the unchanged condition. For the great apes and humans, the unaltered or ancestral skull morphology was assumed by Bonde to be like that of the great apes, with humans derived from it by paedomorphosis, despite the fact that human and gibbon skull form show many similarities. Bonde assumed that the three great apes are not monophyletic and that chimpanzees are more closely related to humans than to gorillas, and therefore that if the ancestral condition was similar to humans and gibbons the great ape morphology would have evolved three separate times in parallel. By Bonde's model, therefore, it was more parsimonious to postulate the great ape condition as the ancestral one from which humans were derived by paedomorphosis, returning to something very like the hominoid ancestral pattern.

It is questionable whether there is any such thing in this respect as a great ape pattern as assumed by Bonde (1984). Cranial base growth is similar in the great apes (Dean & Wood, 1984), but there are differences in facial growth in the orang utan (Giles, 1956), and the human skull differs from both by a combination of increases and decreases in growth rates superimposed on differences in size and shape present at birth. No single heterochronic change underlies the adult shape differences (Dean & Wood, 1984), so that there is no simple ontogenetic change by which human cranial morphology could be derived from that of the great apes, which in any case differ between themselves (Giles, 1956). There is therefore no evidence to support Bonde's (1984) interpretation of the great ape and human ancestral state and no obvious ontogenetic polarity to suggest what that might be.

Chromosome structural morphology

Chromosome structure has been a useful taxonomic tool in some groups of animals and more especially plants, but paradoxically, perhaps, the genetic control of chromosome structure is not well understood. Cytogenetic evidence has been used for primates by several groups of workers recently, but unfortunately both the evidence that they use and the way in which they interpret it have varied considerably (Yunis & Prakash, 1982; Marks, 1983). Chromosomal variation within a clade is large relative to genetic distance or morphological diversity (Marks, 1983), and this causes some of the problems in interpretation. In addition, gibbon karyotypes are highly derived and appear to be evolving more rapidly than those of the great apes and humans, so it is difficult to use them for outgroup comparisons. Changes within the gibbons and within the great ape and human clade consist mainly of pericentric inversions (Mai, 1983; de Boer & Seuanez, 1982), which produce greater proportions of metacentric chromosomes without changing the diploid number and details of banding on the chromosome arms. Based entirely on this evidence, but emphasizing different aspects of it, Miller (1977) argued that humans and gorillas were most closely related, Yunis & Prakash (1982) that humans and chimpanzees were most closely related, and Marks (1983) that chimpanzees and gorillas were most closely related. There appears to be no good reason why chromosomes cannot be used like any other morphological system for inferring relationship, and also no good reason why they should be of any greater importance than other structural systems, but it appears that their full potential has yet to be realized.

Blood group morphology

By contrast with chromosomes, the genetic control of blood group synthesis is well understood in some cases. Some of the blood groups appear to be genetically simple and phylogenetically stable (Socha & Ruffié, 1983), particularly the ABO system, which has not changed substantially from the time of origin of the catarrhines. The major distinguishing feature of the hominoids in the ABO system is the presence of antigens on the red blood cells, whereas in other anthropoids they are restricted to the secretions. Within the Hominoidea, differences are in the levels of polymorphisms rather than being absolute differences, and in the future, and provided sufficient numbers of apes can be maintained in the wild, studies in polymorphisms could be as productive as similar studies on living humans have been. This work, however, remains to be done, and so far there is little useful taxonomic information available.

Protein distance data

The chromosome and blood group evidence is unquestionably structural; that is, it can be treated in exactly the same manner as morphological evidence. With protein structure, the distinction between structural and genetic evidence is not so clear-cut. Whole proteins are studied by a variety of immunological and electrophoretic techniques, and protagonists of protein analysis claim that it provides genetic information directly, a blue-print, as it were, of the DNA molecule, but in reverse. This claim can be justified to the extent that proteins are chains of amino acids, and that changes in bases at the DNA level produce changes in amino acids, these in turn producing changes in the proteins. Substitutions are not entirely on a one-to-one basis, however, for the genetic code is degenerate, and the same amino acid may be coded by different DNA triplets. At the protein level, single amino acid substitutions vary in their effect depending on their location in the protein structure, those in key positions having major effects on three-dimensional structure. Other amino acid substitutions have varying effects on protein structure, but most have little or no effect on the function of the protein, and to this extent most amino acid substitutions appear to be functionally neutral (Perutz, 1983). For instance, despite many changes in the amino acid composition of haemoglobins between hominoids and various mammals, there has been little or no change in function, so that they would appear to have been selectively neutral.

There are a number of criticisms which may be directed at whole protein work. I have recently reviewed these elsewhere (Andrews, 1985*b*) and will comment on them only briefly here. Immunological techniques measure degrees of antigenic difference resulting in an index of dissimilarity (ID). The logarithm of ID was claimed by Sarich & Wilson (1967; Sarich & Cronin, 1976) to have a straight-line relationship with time, but this has been shown to be incorrect (Read, 1975; Corruccini *et al.*, 1980; Gingerich, in prep.; Andrews, 1985*b*). It has also been demonstrated that the relationship of ID with other molecular evidence, such as DNA–DNA hybridization distances, is also non-linear, so that one or both cannot vary at a constant rate with time (Corruccini *et al.*, 1980; Ruvolo, 1983; Ruvolo & Pilbeam, 1985). In addition, it has been shown that there is no empirical or theoretical justification for use of the exponential function in this instance (Read & Lestrel, 1970), that different clustering techniques produce rates of change varying by a factor of up to three (Ruvolo, 1983), and even the experimental techniques have been questioned (Friday, 1981). It has been demonstrated by Goodman (1981; Goodman *et al.*, 1983*b*) in his work on amino acid sequencing that rates of change vary between different proteins and within certain proteins in different lineages, and these variations account for the difficulties encountered in protein immunology (and electrophoresis – Bruce & Ayala, 1979). It must be concluded, therefore, that changes in protein structure vary both non-uniformly with time and in non-linear fashion with other molecular indices; they vary according to the statistical method used to compute the time relation or the clustering method to group the taxa; and they vary also within and between different lineages.

These criticisms of whole protein analysis have been directed essentially at experimental procedure. They can be put aside for the moment to try to understand exactly what information is being provided by protein immunology, assuming constancy of change and an agreed method of measuring the change. Immunology provides a similarity coefficient based on degree of antigenic similarity, expressed as immunological distance. No information is provided on what parts of the proteins are similar, and so it cannot be known whether they are homologous or not. Homoplasy would also contribute to protein similarity, and only if all change were divergent would the effects of this be excluded. Further, the direction of change is unknown. The distance statistic resulting from protein comparisons is computed additively, so that similarity or difference cannot be interpreted cladistically. Take, for instance, one species that has an above average number of uniquely derived changes compared with two other species retaining the ancestral condition for the same characters: on the additive model the two similar species would be linked despite (or because of) the fact that they share only plesiomorphous characters, while the single species with many autapomorphous characters would be made the outgroup. Thus, protein distances are strictly phenetic, and attempts to describe them as cladistic (Sarich & Cronin, 1976) are misplaced.

Put more generally, no distance data of any sort can be used for phylogenetic reconstruction unless the underlying assumptions can be confirmed independently (Farris, 1981). Constancy of rate can be assumed only if it can be checked independently with a time-scale unrelated to the primary source of data. Protein rates cannot be shown to be constant if constancy itself is assumed to calculate the rates.

Amino acid sequencing

It is possible to identify the actual changes that contribute to immunological distance by determining the sequence of amino acids for proteins. Sequencing eliminates the technical problems discussed above for whole protein analysis, but the problems of homology and polarity still remain.

It is difficult at the level of sequence analysis to distinguish homology from homoplasy, even though the actual amino acid changes can be inferred. Probability models have been developed to get around this problem.

Parsimony analysis (Felsenstein, 1981, 1983) in itself cannot make this distinction, and in fact it assumes that most change is divergent. A parsimony solution can be arrived at by choosing among all possible trees (sets of relationships) the one that has the least amount of homoplasy. (Homoplasy in this sense refers to character states shared by taxa but not by their last common ancestor.) Maximum likelihood (Bishop & Friday, 1985) solves the problem in a similar way but by computing all possible branching points progressively within a tree, starting from a 'big bang' assumption of equal relationship between all contained taxa (Bishop & Friday, 1985).

The determination of polarity is possible in some cases of amino acid change. Outgroup comparison is applied in the same way as for morphological comparisons, but there is the added possibility that, by examining the DNA base substitutions required for amino acid substitutions, some indication of direction of change can be inferred. This has been done, for example at position 110 of the myoglobin amino acid chain (Romero-Herrera *et al.*, 1976) for which it has been shown that two progressive substitutions occurred in the Anthropoidea. Old World monkeys and apes have two different amino acid residues at this position, and both are different from the presumed anthropoid ancestral condition; the most parsimonious solution would have had the two groups changing independently from the ancestral condition, but in terms of DNA substitutions this would have required one extra change, because two substitutions are needed for the change from alanine (ancestral anthropoid) to cysteine (hominoid). One of these substitutions could have produced the Old World monkey residue (serine), and so the most parsimonious solution at the DNA level is the assumption of an intermediate stage during the change to the hominoid condition in which the ancestral catarrhine condition was equivalent to that of living Old World monkeys (Andrews, 1985*b*).

Amino acid sequencing has provided some resolution of hominoid relationships, but the finer details do not appear amenable to this approach because there are so few differences between closely related species (Goodman, Baba & Darga, 1983*a*; Bishop & Friday, 1985).

DNA molecular evidence

The final stage in this brief review of evidence is the analysis of the DNA molecule itself. The changes in DNA with which I will be mainly concerned here occur by point mutations, which consist of substitutions of one base for another within one of the 64 triplets that make up the genetic code. Transitions greatly predominate over transversions (see below for discussion of these terms, and Brown *et al.*, 1982; Templeton, 1983; Hasegawa, Yano & Kishino, 1984) and the level of homoplasy is very high when both are counted together.

Three types of analysis have been published recently: DNA–DNA hybridization (Sibley & Ahlquist, 1984); restriction endonuclease site mapping (Ferris, Wilson & Brown, 1981*a*; Ferris *et al.*, 1981*b*; Brown *et al.*, 1982; Templeton, 1983); and DNA sequencing (Chang & Slightom, 1984; Goodman *et al.*, 1984; Harris *et al.*, 1984; Hasegawa *et al.*, 1984).

DNA–DNA hybridization produces a distance statistic based on closeness of fit for pairs of species (see Sibley & Ahlquist, this volume, pp. 97–100). It measures degree of DNA sequence correspondence based on proportions of identical to non-identical sequences, and to do this it assumes that all sequence similarity is homologous. Identity between bases in any one part of the sequences does not of itself indicate homology, for homology depends on the inheritance of any given base change from a common ancestor, but if enough of the bases are similar in two sequences to enable reassociation to occur, and given the length and complexity of the sequence structure, it is most unlikely that sequence identity could be achieved by any means other than through shared descent, i.e. homology. This is exactly the same situation as for any structural character: the greater the complexity of shared characters, the less likely is this to be due to homoplasy. The problem with DNA–DNA hybridization is the same as for any distance statistic, that it is not possible to isolate the changes that are being measured, and so it is inherently untestable.

Similar characters, be they complex morphological characters or nucleotide substitutions, may be inherited unchanged from a remote common ancestor, in which case they are primitive retentions, or they may be inherited from the last common ancestor, in which case they are derived. The small number of derived characters uniquely shared between two closely related species would be swamped by the much larger number of shared primitive characters unless the latter can be eliminated from consideration, but this is exactly what has not been done by the hybridization method.

Sequencing the DNA molecule is still at an early stage of development, and it cannot yet compare in scope with the hybridization method which covers the entire genome. However, the short segments of DNA that have been sequenced provide much more precise information than distance methods, and in the long term it is undoubtedly sequencing that will provide the answers. DNA sequencing provides information like that from amino acid sequencing, with many of the same problems for phylogenetic reconstruction. Convergence is minimized by parsimony methods, but again there is no theoretical or empirical way of distinguishing convergent change from homologous other than by probability.

Early DNA sequencing work was done on hominoid mitochondrial DNA (mtDNA), which has much higher rates of change than nuclear DNA (Brown *et al.*, 1982; Miyata *et al.*, 1982; Hasegawa *et al.*, 1984). This leads to problems in phylogeny reconstruction if different types of change, such as transitions and transversions, are not distinguished. Transitions are changes within the purine and pyrimidine pairs (adenine and guanine, thymine and cytosine, respectively), while transversions are changes between them. Transitions accumulate so rapidly in mtDNA that within about 20 Myr they have 'saturated' all readily substituted positions and further change results in back mutations. Change continues in a backwards and forwards manner, one position changing many times, and it is not possible to detect this by analysis of either sequence or DNA–DNA hybridization data. Brown *et al.* (1982) recognized this problem and concluded that the periods of time for which mtDNA divergence gave meaningful results is relatively short (no more than 20 Myr). An alternative was proposed by Hasegawa *et al.* (1984), that only transversions should be counted; these occur much less frequently (by a factor of 8) and are correspondingly less likely to be reversed.

The work on mtDNA has recently been shown to be suspect because of interspecific transfer of mtDNA (Ferris *et al.*, 1983; Spolsky & Uzzell, 1984). The mitochondria are maternally inherited, and if any hybridization occurs across species boundaries, the mitochondria from the female may become established and even replace those of the population from which the male was derived. In the long term the effects of this are probably minimal, but for distinguishing between closely related species, it could be of very great importance.

It has also been shown recently that the rates of change in DNA may alter. There are well-established differences between rates of silent or synonymous substitutions and non-synonymous or replacement substitutions, and also between rates in mtDNA and nuclear DNA. Wu & Li (1985) have shown that nucleotide substitutions are more frequent in rodents than in man, synonymous substitutions by a factor of 1.3 and non-synonymous substitutions by a factor of 2. It has been claimed for DNA mutation rates that they are constant over time and for different lineages, and variations by up to a factor of 2 would greatly alter the conclusions of both the hybridization and sequencing methods, as they affect the relative timing of the events, although not, of course, the order of branching.

This completes the review of types of evidence, from morphology to DNA, and this evidence will now be applied to two branching points in hominoid evolution.

Early hominoid branching points

There is little doubt that the earliest branching point in the hominoid radiation was that between the gibbons and the rest of the hominoids (the great apes and man). In many aspects of their cranial morphology and dentition the gibbons are close to the ancestral hominoid morphotype, but in their postcranial and cytogenetic morphology, and probably also in their behaviour, they are the most specialized hominoid group (Andrews & Groves, 1975; Mai, 1983; Marks, 1983; Andrews, 1985*a*). This mixture of primitive and autapomorphous character states makes the gibbons an easily defined group and one also that is useful for outgroup comparisons with the other hominoids.

The relationships of the great ape and human clade are less easy to ascertain. Four of the possible sets of relationships are shown in Fig. 2.2, and the evidence for each of these will be considered in turn. The first gives what is called the traditional view based on morphology, showing the great apes as a whole distinct from humans. There is some morphological evidence supporting this distinction (Kluge, 1983), and it has been claimed that there is also some cytogenetic evidence supporting it, but there is reason to doubt this (p. 27 above).

The second set of relationships links humans with the orang utan (Schwartz, 1984). Again, both morphological and cytogenetic evidence (Miller, 1977) has been advanced in support of this view. The third and

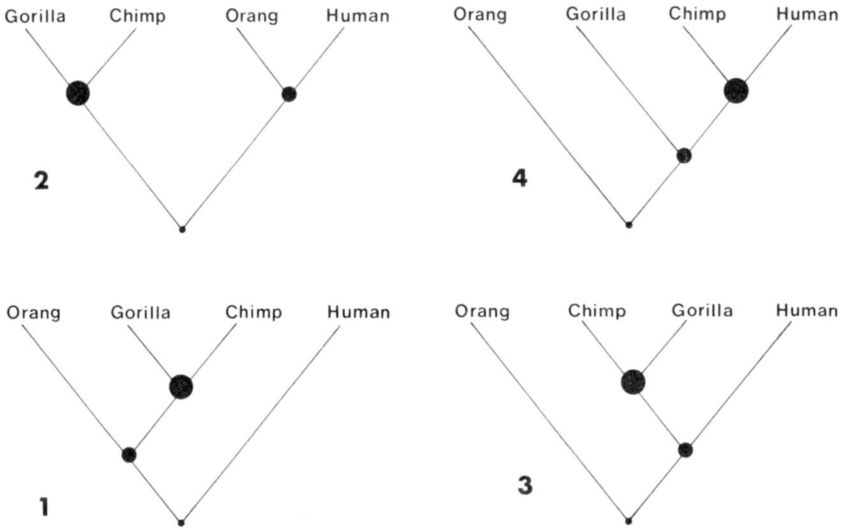

Fig. 2.2 Four cladograms showing four of the possible sets of relationships of the great apes and humans. 1 shows the three great apes as a clade, with man as the outgroup; 2 shows the orang utan related to humans, with the African apes as the outgroup; 3 shows an African ape clade linked with humans, with the orang utan as outgroup; and 4 shows humans and chimpanzees linked, with the gorilla as the outgroup, and the orang utan as the outgroup to them all.

Table 2.1. *Synapomorphies of the orang utan clade*

Morphological character state present uniquely in the orang utan clade	Equivalent character state present in the African apes, humans and gibbons
Orbits higher than broad	Orbits square or broader than high
Interorbital pillar narrow	Interorbital pillar broad
Reduced glabella	Swollen glabella
Zygomatic foramina numerous, above level of inferior margin of orbit	Zygomatic foramen single, below orbit
Palatine foramen small, slit-like	Palatine foramen larger
No incisive fossa	Deep incisive fossa[1], incisive fossa extends through palate[2]
Nasal floor unstepped	Nasal floor stepped
Ulnar styloid process lacks contact with scaphoid	Ulnar styloid process contacts scaphoid
Incisors: I1 > I2 (150–200%)	I1 > I2 (110–130%)
Molars with low dentine horns	Molars with high dentine horns
Enamel partly pattern 1, intermediate and slow formed	Enamel partly pattern 1, slowly formed[3] enamel pattern 3[4]
Molars with deep secondary wrinkling	Enamel smooth or showing slight wrinkling
Processus vaginalis persists	Processus vaginalis obliterated
Large pars intermedia of adenohypophysis	Absent[1], weak[2]
Processus infundibuli present	Absent
Small axillary organ	Large axillary organ[1], absent[2]
Sparse hair on scalp: 160 cm^{-2}	More abundant hair (185–330 cm^{-2})[1]
Filiform tongue papillae restricted to middle of tongue	Tongue papillae extend to sides of tongue
Fungiform tongue papillae few	Fungiform papillae many
Small gall bladder, lacking bend	Large gall bladder, with bend

[1] African apes and man only
[2] Gibbons and other catarrhines
[3] African apes only
[4] Gibbons and humans

fourth cladograms are the ones usually associated with the molecular evidence, with the African apes related to humans and the orang utan as the outgroup. There is also abundant morphological and cytogenetic evidence which supports this, much of it already published (Cave & Haines, 1940; Mai, 1983; Marks, 1983; Andrews, 1985*b*; Martin, 1985*a*, *b*).

One possible source of confusion, which can be dealt with immediately, is the specializations unique to the orang utan. Characters unique to this species are most likely to be autapomorphies, and the alternative character states common to the gibbons, African apes and humans are symplesiomorphies. These probable autapomorphies are listed in Table 2.1, together with the alternative character states present in other hominoids. In most cases the alternative conditions are also present in catarrhines generally and are almost certainly ancestral hominoid characters. In this listing, which is by no means exhaustive, there are eight skeletal characters, four dental, and eight reproductive and visceral, making twenty in all.

Great ape clade

In 1983, Kluge described 11 characters of the reproductive system, musculature, skeleton and skeletal proportions that were shared by the orang utan, chimpanzee and gorilla. These formed the basis for his recognition of a great ape clade, for he attempted to show that the characters were probably derived with respect to the common ancestor of humans and great apes. The main basis for the recognition of sister-group relationship among the great apes is their overall similarity based on their broad flattened thorax and reduced vertebral column length, together with their elongated arms and hands and reduced legs. Morphological correlates of the thorax shape are the dorsal position of the scapula, the medial rotation of the humeral head, and the lengthened dorsal spines of the cervical vertebrae (related to the more cranial origin of the stabilizing muscles of the shoulder). The reduction of the lumbar vertebrae to four is related to but not the sole cause of vertebral column shortening, and reduction in length of the thumb (relative to the hand or forelimb length) is the correlate of limb lengthening. All of these characters are, in fact, directly related to increase in body size: the allometric relations for thorax broadening (Andrews & Groves, 1975) and forelimb lengthening (Aiello, 1981) to body size determine the development of these characters in all of the hominoids except gibbons, which are highly derived for these features. There is one exception to this, and that is the reduction in length of the hindlimbs of the great apes; all of them fall significantly below the catarrhine allometric gradient for the relation of this character to body size, and it might be considered a great ape synapomorphy, but it has recently been shown (Aiello & Day, 1982) that early hominids also had this hindlimb reduction (combined with clear evidence of bipedal gait), so that it appears more likely to be a great ape and human character.

The correlation of these morphological characters with body size reduces the eleven great ape characters described by Kluge (1983) to six. Of these, the two characters of the reproductive system appear to be correct, although of limited phylogenetic value. The separation of the flexor tendons of the toes is a strong character, for it does not seem to be correlated in any simple way to arboreal locomotion, for equally arboreal gibbons and Old World monkeys lack this character. The attenuation of flexor pollicis longus, on the other hand, is less easy to assess, for although it occurs in all of the great apes, in the orang utan the thumb flexors are reinforced by two accessory short flexors. Also, the thumb is not reduced in relation to body weight in the orang utan. Finally, the prognathism of the face, said to be a synapomorphy of the great apes by Kluge, appears to be quite distinct in the orang utan compared with the African apes and thus was probably developed in parallel. The simian shelf, another supposed synapomorphy, was shown some time ago (Delson & Andrews, 1975) to be primitive for the Hominoidea.

The conclusion that must be reached on cladogram 1 of Fig. 2.2 is that, although there are a number of similarities shared by the great apes, they are not convincing evidence of phylogenetic relationship. They are a mixture of primitive retentions, parallelisms or allometric correlates of body size, and when these have been eliminated there are just three characters, two of the reproductive system and the separation of the toe flexor tendons, that support cladistic relationship between the great apes.

Human and orang utan clade

It has recently been suggested that there is evidence for relationship between orang utans and humans (Schwartz, 1984). Schwartz listed 15 uniquely shared features as evidence of relationship. Some of these appear to be of doubtful significance – for instance, the records for longest hair or longest copulation bouts – and it is difficult to evaluate some of the other characters – for instance, the widespread mammary glands or the female genitalia lacking tumescence – without further work. Thick enamel is said by Schwartz to be shared by humans and orang utans, but it has been shown (Martin, 1983, 1985*a*) that orang utans do not have enamel either as thick as humans or with the same microstructure, so that there is no evidence of homology. It has also been shown that thick enamel may be primitive for the great ape and human clade (Martin, 1985*a*). Another of Schwartz's characters, low-cusped cheek teeth, is a direct consequence of thickened enamel, for the thickening tends to diminish cusp relief, and it cannot be treated as an independent character.

Another character that has been well studied is the morphology of the incisive canal in hominoids (Ward & Pilbeam, 1983) (Fig. 2.3). In the orang utan there is no incisive fossa and the incisive canal is long, narrow and slit-like. In humans there is a deep incisive fossa that is transversely broad and divided into two chambers. The incisive canal is larger in diameter and shorter, and it carries both nerves and a blood supply to the palate. In this respect the human condition is different from the orang utan condition and is identical to that of the African apes, and it is likely that this morphology is primitive for the great ape and human clade. The monkeys and gibbons are different, with no canal developed and a large foramen connecting the nasal passage to the palate, and this would seem likely to represent the ancestral hominoid pattern from which the African ape and human morphology was derived. This ancestral morphology is also seen in the fossil apes *Dryopithecus* and *Proconsul*. From this was further derived the orang utan pattern, which is present in another fossil ape, *Sivapithecus* (Andrews & Cronin, 1982; Ward & Pilbeam, 1983).

Other characters described by Schwartz (1984) include the presence of reduced lower third molars in the orang utan and humans. This is correct, but it is a character also of chimpanzees and gibbons and so it is likely to be an ancestral character of the hominoid clade. The same is true for the lack of upper molar cingula. Some of the characters of the reproductive system also appear relatively insignificant. The gestation time is similar in orang utans and

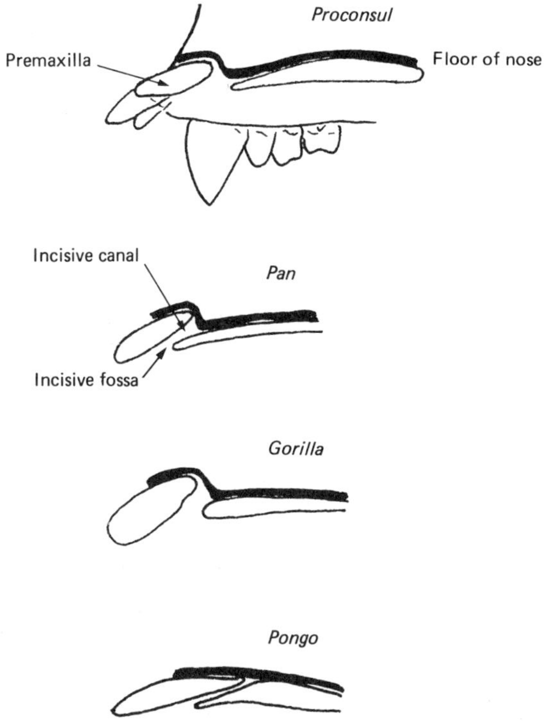

Fig. 2.3 Sagittal sections of the nose region. The ancestral hominoid condition is represented by the Miocene *Proconsul* at the top, and this condition is seen also in gibbons and the Miocene *Dryopithecus*. The African ape and human condition is represented here by chimpanzee and gorilla. The orang utan is shown at the bottom, and note the rotated premaxilla and flattened floor of the nose (after Ward & Pilbeam, 1983).

humans, slightly longer than in gorillas (Cross & Martin, 1981) and considerably longer than in chimpanzees. These differences have a different perspective when gestation time is corrected for body size. After calculating the mammalian allometric relation of gestation time to body size (Martin & MacLarnon, 1985), the residual distances from the line show the orang utan to be most divergent (0.36), chimpanzees and gorillas least divergent (0.31), and humans and gibbons intermediate. The similarity in gestation time therefore appears to be of doubtful significance in humans and orang utans. Continuous receptivity of females during the oestrus cycle is least restricted in the orang utan and most restricted in gorillas, with chimpanzees in between, but in no species does it reach human levels, so that again this is not a secure character. When these poorly defined and/or primitive characters are removed from consideration, there remain four putative synapomorphies and another two or three possibles.

The conclusion on cladogram 2 of Fig. 2.2 is therefore that the evidence supporting cladistic relationship between humans and orang utans is not nearly as strong as claimed by Schwartz (1984). There are, however, Schwartz's (1984) putative synapomorphies which have to be considered, including the presence of a foramen lacerum basicranially, delayed ossification of epiphyses, and infrequent and unkeratinized ischial callosities.

African ape and human clade

The remaining two cladograms in Fig. 2.2 show the orang utan as outgroup and the African apes and humans more closely related to each other. For the present discussion the latter two can therefore be considered together. The autapomorphous characters of the orang utan have already been listed (Table 2.1), and if this interpretation of character polarity is correct, the alternative state of these 20 characters cannot be used as evidence of relationship between the African apes and humans. There are twelve additional characters listed in Table 2.2 that appear to be synapomorphies of chimpanzees, gorillas and humans, and they include, from top to bottom, four cranial characters, one postcranial, two characters of the reproductive system, and five soft tissue characters.

The African apes and man also share the cytogenetic character of having areas of Q-brilliance on a number of their chromosomes, unknown in other mammals (Marks, 1983). On the Y chromosome, the brilliance is confined to gorillas and humans, and this has been put forward as a synapomorphy demonstrating cladistic relationship between these two species (Miller, 1977), but Marks (1983) considers that, since this is such a rare character in mammals, it is more likely that Q-brilliance was present on the Y

Table 2.2. *Synapomorphies of the African ape and human clade*

Morphological character state present in the African apes and humans	Condition in other hominoids
Frontal sinus present	Absent
Supraorbital tori developed, continuous	Absent
Postorbital sulcus developed	Absent
Greater middle ear depth (8.5–12.5 mm)	Less: orang utan 6.8 mm, monkeys 5.0–8.3 mm
Fusion of os centrale to scaphoid	Not fused except occasionally in orang utans
Subdivision of the prostate	None
Large uterus	Small
Apocrine glands sparsely distributed over body	More abundant
Eccrine glands abundant over body	Few in monkeys, more abundant in the orang utan
Large axillary organ	Small in orang utan, none in others
Small single larynx tuberculum cuneiforme	Large
Low proportion (3–21%) type I aorta	High proportion (63–100%)

chromosome, along with the others, in the common ancestor of the African apes and humans and was secondarily lost in chimpanzees. In view of the lack of supporting evidence for a gorilla–human relationship, this seems the most reasonable interpretation. The alternative interpretation that Y-brilliance is convergent in gorillas and humans is put forward by Mai (1983), but the same argument applies, that such an unusual character would be unlikely to evolve twice. There is no shortage of different interpretations among cytogeneticists even when based on the same evidence!

Blood group polymorphisms are not particularly useful at this taxonomic level. There is a suggestion in the MN system that the shared synthesis of both M and N in the African apes and humans may be a synapomorphy, even though there are differences at the gene level: humans have two alleles, *M* and *N*, which control the synthesis of the corresponding factor to give M, N and MN; chimpanzees have a single allele, *M*, which controls the synthesis of M and MN (but no N); gorillas have a single allele, *N* which controls the synthesis of N and MN (but no M) (Socha & Ruffié, 1983). Orang utans can synthesize only M.

Protein immunological results demonstrate greater degrees of similarity between the African apes and humans (Goodman, 1963, 1973, 1974, 1975, 1976; Sarich & Wilson, 1967; Dene, Goodman & Prychodko, 1976; Sarich & Cronin, 1976; Baba, Darga & Goodman, 1980; Goodman *et al.*, 1983*a*, *b*).

These similarities are usually expressed as a measure of immunological distance, which varies irregularly with time, so that, although immunology lends support for cladograms 3 and 4, not much reliance can be placed on this.

By determining the sequence of amino acids in proteins, the nature of the similarities that go to make up the immunological distances can be identified. In the myoglobin amino acid sequence, the African apes and humans share one residue change at position 23 to glycine, whereas other hominoids, and catarrhines generally, have serine at this position (Romero-Herrera *et al.*, 1976, 1978). There is good reason, therefore, to see this as an African ape and human synapomorphy, with the orang utan as outgroup. In the haemoglobins, the African apes and humans also differ from the orang utan by another two residues: β87 threonine instead of lysine, and β125 proline instead of glutamine (Goodman *et al.*, 1983*b*). In the first case, it is not clear which of these alternatives is more likely to represent the ancestral condition, but in the second, the orang utan glutamine at position β125 is shared with all other catarrhines. Doolittle *et al.* (1971) showed that two substitutions are shared by the African apes and humans in the fibrinopeptides, and while these are claimed to be synapomorphies for this clade, no evidence is put forward to support this view. Chimpanzees and humans have also been shown to differ from orang utans at five loci in the carbonic anhydrase amino acid sequence (Tashian *et al.*, 1976), but at least three of these can be shown to be uniquely derived for the orang utan and therefore primitively retained for the others. It is possible that the remaining two could be African ape and human synapomorphies, but it would be interesting to know the condition in gorillas.

Evidence from the DNA is unequivocal in placing the African apes and humans together and the orang utan as the nearest outgroup (Brown *et al.*, 1982; Hasegawa *et al.*, 1984; Sibley & Ahlquist, 1984). The evidence from this source will be considered later.

Conclusion

In the descriptions of the morphological evidence, it has been shown that there is some evidence supporting all four of the cladograms in Fig. 2.2. After elimination of doubtful characters, there are three characters which support the great ape clade (chimpanzee, gorilla and orang utan), with humans as the nearest outgroup; there are four characters which support the orang utan and human clade, with the African apes as the nearest outgroup (the evidence for a chimpanzee–gorilla clade will be discussed in the next section); and there are twelve characters which support the African ape and human clade, with the orang utan as the nearest outgroup. The morphological evidence thus tends to support the last alternative (cladograms 3 and 4 in Fig. 2.2) but there is clearly some homoplasy to be accounted for.

The distinction between an African ape and human clade and the orang utan is also indicated by the cytogenetic and blood group evidence, although this evidence does not appear to be as clear as it might be. The same applies to immunological and electrophoretic analyses of proteins and DNA–DNA hybridization, which provide distance statistics supporting this hypothesis. Amino acid sequencing demonstrates the presence of several uniquely shared amino acid residues in the African apes and humans, and DNA sequencing shows that they share the greatest number of DNA substitutions.

It can be concluded, therefore, that nearly all evidence supports the branching points shown on cladograms 3 and 4 of Fig. 2.2, that is the existence of an African ape and human clade with the orang utan as the nearest outgroup. In the next section, the branching points within this clade will be considered.

Human–ape divergence

There are three possible sets of relationships for the species within the African ape and human clade. Chimpanzees and gorillas can be linked, with humans as the outgroup (Fig. 2.2, cladogram 3); chimpanzees can be linked with humans, gorillas being the outgroup (Fig. 2.2, cladogram 4); or gorillas and humans can be linked, but there is so little evidence supporting this that it is not considered further here. The other two can be expressed in anthropocentric terms as the choice between whether our last common ancestor was with chimpanzees and gorillas together or with chimpanzees alone. The predicted characteristics of this last common ancestor depend on the correct identification of these relationships.

Chimpanzee–gorilla clade

There are a number of details of morphology by which the African apes resemble each other and differ from humans and other hominoids. Two of these, the enamel pattern and the complex of characters associated with knuckle-walking, are of some significance, and in addition there are a further three characters. These are listed in Table 2.3.

The recent work of Martin (1983, 1985*b*) has shown that the ancestral great ape and human condition for molar enamel is thick pattern 3 enamel. This is retained by living and fossil humans, but in chimpanzees and gorillas the

Table 2.3. *Synapomorphies of the African apes*

Morphological character state shared uniquely by chimpanzees and gorillas	Condition in other hominoids
Six sacral vertebrae	Five or less
Short ethmoid-lacrymal contact (40–90%)	Long contact (100%)
Fronto-maxillary contact in orbits (30–50%)	No contact
Enamel thickness and structure	
Thin enamel	Variable but thick in man and in the inferred great ape and man ancestral state
40% pattern 1 enamel	Pattern 3 (except orang utan – see text)
Enamel accretion rate < 1.5 μm/day	5–7 μm/day (see text)
Knuckle-walking adaptations	
Dorsal transverse ridges on metacarpal leads	Absent
Dorsal extension of articular surface of metacarpals	Absent
Dorsal ridges on distal radius and scaphoid	Absent
Volnar and ulnar inclination of distal articular surface of the radius	Absent
Well-developed trochlear ridge of humerus	Trochlear ridge not prominent
Very deep olecranon fossa of humerus	Olecranon fossa not as deep
Thumb short relative to body weight	Thumb longer
Flexor digitorum superficialis strongly developed	Not so strongly developed
Shortened flexor tendons, so that fingers cannot be straightened when hand is dorsiflexed	Flexor tendons not shortened
Knuckle pads over middle digits	Absent

enamel is secondarily thinned by a change of prism structure from pattern 3 to pattern 1. Pattern 1 enamel in these species is formed at a rate of about 1.5 μm per day (the size of the cross striation repeat interval), as compared with the pattern 3 rate of 5–7 μm per day in human enamel. The formation time is approximately the same, but because it forms so much more slowly in the African apes, the enamel is much thinner, the same thickness as gibbon enamel, in fact, but in gibbons the enamel is all fast-formed pattern 3 enamel (the ancestral hominoid condition) formed over a shorter period. The orang utan also has some pattern 1 enamel, but less than in the African apes (20 per cent as opposed to 40 per cent) and formed in a different way: 200 μm at 2.5 μm per day and a further 50 μm at 1.8 μm per day. The possibilities of reducing enamel thickness by this means appear considerable, and the chances of two species arriving independently at the same reduction rates must be very small. For this reason, therefore, the similarity in chimpanzee and gorilla enamel seen in both the method and rate of reduction and in the end result is considered to be highly significant.

The same argument applies to the larger complex of characters related to knuckle-walking. This is a highly unusual form of quadrupedal locomotion, not practised by the orang utan, despite its similarity in size and gait to chimpanzees and gorillas. There is no evidence for any adaptation towards knuckle-walking in any known human fossil material, for which good evidence extends back to just over 3 MyrBP. It therefore appears to be unique for the African apes, but the question remains whether this form of locomotion could have evolved in parallel in chimpanzees and gorillas; in other words, is it in any way a necessary adaptation for large-size/closed-hand locomotion as practised by the great apes? Table 2.3 shows a complex of 10 characters of the elbow, wrist and hand that are either related to stability of these regions or are directly connected with knuckle-walking, and for all of these the chimpanzee and gorilla show identical adaptations (Tuttle, 1967, 1969). None of these characters is present in the orang utan, even though it has a similar gait when on the ground (i.e. large body size combined with closed-hand walking). The orang utan supports itself on the sides of its clenched fists rather than on its knuckles, and in locomotor terms this appears to be a close parallel to knuckle-walking proper, but in morphological adaptations it is quite distinct. This suggests that, as in the example given above for tooth enamel, the orang utan has arrived at a similar adaptation but with different characters, whereas the chimpanzee and gorilla have arrived at the same end result by identical means, suggesting that in the latter their shared characters are homologous.

The other morphological characters listed in Table 2.3 are of uncertain value. They certainly appear distinct for the two African apes, and they differ from the condition in other catarrhine primates, but whether they have any phylogenetic significance is difficult to say. There are some cytogenetic characters that support a link between chimpanzees and gorillas (Stanyon & Chiarelli, 1982), but no evidence from study of the blood groups. Protein immunology cannot resolve the trichotomy between the chimpanzee, gorilla and human (Andrews & Cronin, 1982), and there are no shared amino acid substitutions between the two African apes. There is, however, some evidence from DNA sequencing that suggests a chimpanzee–gorilla grouping (Brown *et al.*, 1982; Templeton, 1983), but these are better discussed in the next section, since other interpretations of the same or similar data show, on the contrary, that humans and chimpanzees are more closely related.

In conclusion, therefore, there is very substantial morphological evidence supporting a link between chimpanzees and gorillas (cladogram 3 of Fig. 2.2). This is supported by some interpretations of the DNA sequence evidence, discussed more fully below.

Human–chimpanzee clade

In contrast to all of the possible cladistic groupings so far discussed, there is no morphological evidence whatever supporting a cladistic relationship between humans and chimpanzees. There is also no good evidence from studies of blood groups and chromosomes. The RLEF blood group system in hominoids has been shown to be equivalent to the Rhesus system in humans (Socha & Ruffié, 1983), and the chimpanzee with 24 RLEF types is closest to humans, but the pygmy chimpanzee (*Pan paniscus*) is quite different from the common chimpanzee (*Pan troglodytes*) in this respect: the pygmy chimpanzee has only one RLEF type, $R_{ab}CE$, a rare variant in other hominoids that is completely absent in the common chimpanzee. This makes it hard to interpret the similarity between chimpanzee and human as homologous.

Humans and chimpanzees share the greatest number of identical chromosome pairs – 13 (Yunis & Prakash, 1982) – but this is only marginally above the number shared by chimpanzees and gorillas. There are no clear chromosomal homologies shared by chimpanzees and humans, but Yunis & Prakash (1982) suggest a complex sequence of changes on chromosomes 2p, 7 and 9 by which they attempt to show that humans and chimpanzees are more similar. Chromosome 7, for instance, was thought to have an ancestral condition like that of the gorilla because the gorilla condition differed from the orang utan by one pericentric inversion and from chimpanzee and human by another, so that the gorilla is somehow intermediate. On the data supplied by Yunis & Prakash, it would be equally possible for the ancestral condition to be orang-like, with the African apes and humans differing by a pericentric inversion and the gorilla by a unique paracentric inversion. This seems to fit their Fig. 2 better than their own interpretation, and it would leave humans and chimpanzees retaining the ancestral hominoid condition.

On chromosome 9 it does seem more likely that the ancestral condition was paracentric (White, 1973), as in orang utans and gorillas, so that it was changed by pericentric inversion to metacentric, shared by chimpanzees and humans, but the latter are so different that it is hard to believe that the same pericentric inversion produced them. Also, on chromosome 2p, for which Yunis & Prakash suggest an orang utan and gorilla-like ancestral condition with chimpanzee derived by a pericentric inversion, it is equally possible for the chimpanzee condition to be primitive. It is simply impossible to tell without adequate outgroup comparison, and this is difficult because the gibbons are so derived cytogenetically.

As has already been mentioned, the immunological and electrophoretic study of proteins is unable to discriminate relations between the species of African apes and humans. This brings us to the sequencing data on amino acids and DNA. There are two amino acid substitutions that support the chimpanzee and human clade (Goodman *et al.*, 1983*b*). At position 23 of the α-haemoglobin chain all other hominoids (and some species of monkey) have aspartic acid whereas chimpanzees and humans have glutamic acid. Similarly, at position 104 on the γ-haemoglobin chain, chimpanzees and humans have lysine, whereas other hominoids and those Old World monkeys that have been tested have arginine. On this evidence there would appear to be a high probability that these two characters are synapomorphies of the chimpanzee and human clade. The myoglobin sequence also shows two differences between chimpanzees and gorillas, but only one between both and humans. This is because each African ape has a unique substitution, and although a simple additive procedure might suggest relationship between humans and one of the African apes (a human-chimpanzee relationship was selected by Romero-Herrera *et al.*, 1978), the presence of unique characters is not evidence of relationship because equal support is provided by these data for a gorilla–human relationship.

Evidence in support of human–chimpanzee relationship has recently been put forward by Sibley & Ahlquist (1984) based on DNA–DNA hybridization results (see also Hoyer *et al.*, 1972; and Sibley & Ahlquist, this volume, pp. 97–100). This produces a distance statistic which fails to distinguish character homology, and this must raise questions about the reliability of the DNA–DNA hybridization results. What they purport to show is greater similarity between chimpanzees and humans (Sibley & Ahlquist, 1984). The range of distances (delta $T_{50}H$) published by Sibley & Ahlquist is 1.3 to 2.3 between humans and chimpanzees, and 1.7 to 2.8 between gorillas and chimpanzees, although Sibley (personal communication) now claims a greater degree of divergence between these two sets of results, with means of 1.6 for the first and 2.2 for the second.

Restriction endonuclease cleavage mapping of mitochondrial DNA (Ferris *et al.*, 1981*a*, *b*) identified a minimum of 147 mutations in the hominoids. Minimum change produced a cladogram linking chimpanzees and gorillas, but with only a single additional mutation, an alternative cladogram linked chimpanzees and humans. Brown *et al.* (1982) obtain a similarly ambiguous result based on sequencing a segment of mtDNA, with trees based on the minimum number of events showing first a chimpanzee–gorilla relationship. With the addition of another two substitutions, chimpanzees and humans could be related, and adding a further one, humans and gorillas.

The ambiguity of the mtDNA results is in part due to the mixing of synonymous and replacement substitutions. It is also in part due to combining transitions and transversions. Brown *et al.* (1982) demonstrated the much more common occurrence of transitions, and showed that within 20 Myr they have saturated all possible substitutions in mtDNA and further changes, which continue at the same rate at the same loci, obliterate the earlier substitutions by back mutations. Transversions, on the other hand, are much less common, and by distinguishing between them, Hasegawa, Kishino & Yano (1984, 1985) have attempted to eliminate this problem. They have added up the numbers of transversion differences observed in the short segment of mtDNA that they analysed, and they have shown that there are five differences between chimpanzees and gorillas, four between humans and gorillas, and only three between humans and chimpanzees. This indicates the least amount of divergence between humans and chimpanzees, in contrast to the results of Brown *et al.* (1982) who analysed exactly the same segment of mtDNA. Brown and his co-workers found that parsimony analysis of the mtDNA sequence slightly prefers the chimpanzee–gorilla association, with 145 events as opposed to 147 for the human–chimpanzee solution. On the other hand, Bishop & Friday (1985) analysed Brown's data by the maximum likelihood method, and found the contrary result, with maximum likelihood for the human–chimpanzee solution: −2916.14 compared with −2920.26 for the chimpanzee–gorilla solution.

An alternative approach to the sequencing data is based on commonality and outgroup comparisons (assuming the orang utan to be the outgroup for the African ape and human clade) so that the probable polarity of change can be determined. This can be arrived at by counting the numbers of shared substitutions that are unique to different pairs of species, where by outgroup comparisons they can be inferred to be different from the ancestral condition for that clade. Figure 2.4 shows the numbers of uniquely shared substitutions, both transitions and transversions, for pairs of species in the African ape and human clade. The largest number (eleven) is shared by humans and chimpanzees; eight are shared by gorillas and chimpanzees, and only two by gorillas and humans. On the assumption that relationship is based on the largest number of shared substitutions, a human–chimpanzee relationship is indicated, so that the eight substitutions shared by chimpanzees and gorillas must be parallelisms. If transitions are distinguished from transversions, all eight of the chimpanzee–gorilla substitutions are found to be transitions, while four of the eleven substitutions shared by humans and chimpanzees are transversions. This appears to provide substantial support for chimpanzee–human relationship.

Conclusion

There is a conflict in the evidence for the relationships within the African ape and human clade. The morphological evidence (Table 2.3) supports cladogram 3 of Fig. 2.2, which shows chimpanzees and gorillas most closely related. There are no features of gross morphology or physiology that are shared by chimpanzees and humans but not by gorillas, whereas there are two major character complexes, hind limb morphology (linked with knuckle-walking) and enamel structure, which are shared by chimpanzees and gorillas. In both cases, homology can be investigated by rigorous delineation of the character states contributing to these character complexes. The probability can also be demonstrated of both being derived characters compared with the great ape and human ancestral morphotype.

Evidence from chromosome structure has also been put forward in support of an African ape clade, but the same evidence has been used by others to support a human and chimpanzee clade. It is not clear at present which of the two is correct.

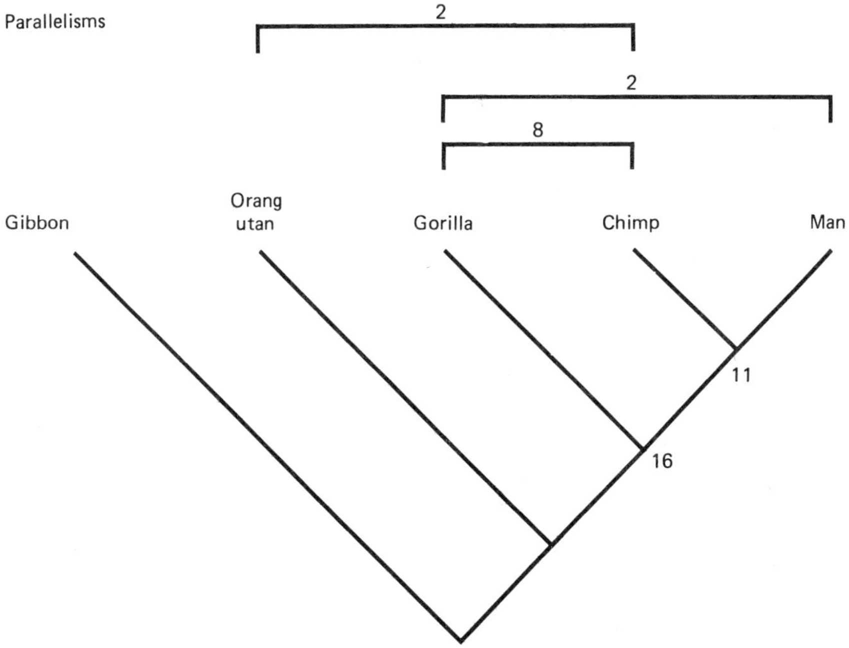

Fig. 2.4 Numbers of exclusively shared substitutions in an 896 nucleotide sequence of mtDNA in the great apes and man (Brown *et al.*, 1982). Only those substitutions exclusively shared by two or three species, where the other hominoids share the outgroup condition, have been counted. The greatest number of substitutions (eleven) is shared by chimpanzee and human; the next highest number (eight) is shared by chimpanzee and gorilla. This provides marginal support for the association of humans and chimpanzees, but what is particularly interesting is the relatively large amount of parallel change that must be accounted for in the genome. These inferred parallelisms are all transitions (see text), whereas four of the eleven shared substitutions in humans and chimpanzees are transversions.

The amino acid sequencing data provide strong evidence in favour of cladogram 4 of Fig. 2.2. Two substitutions in the globin sequences appear to be uniquely derived characters shared by chimpanzees and humans. There is no reason not to believe that they are homologous in chimpanzees and humans, but it is not possible to demonstrate that this is so.

DNA–DNA hybridization data also support the human and chimpanzee clade, but no information is available on the significance of the characters that went to make up this distance statistic. It is confirmed by some of the DNA sequencing evidence, particularly when restricted to mtDNA transversions or when the data are analysed by maximum likelihood. The level of resolution is small in all cases, however, and since other workers have obtained contrary results from the same data base, the conclusions from the DNA appear far from certain.

General conclusion

In the two case studies presented here, the first shows complete congruence between molecular and morphological studies. All lines of evidence show the orang utan to be the outgroup to the African apes and humans (Fig. 2.5) but part of the great ape and human clade. In the second case, however, there is an apparent incongruence. The morphological evidence and some molecular interpretations show chimpanzees and gorillas to be more closely related, while some of the molecular evidence shows humans and chimpanzees to be more closely related. Amino acid sequencing and DNA–DNA hybridization provide the strongest support for the

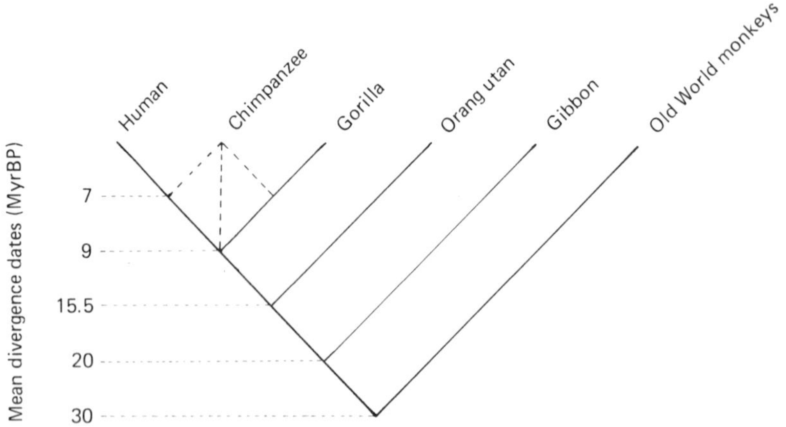

Fig. 2.5 Cladogram showing relationships, and mean divergence dates, of the Hominoidea.

human–chimpanzee link, while gross morphology supports the African ape grouping. DNA sequencing, which ought to provide the answer to this problem, produces ambiguous results: different interpretations of the same data lend support to both alternatives. In view of this, and the strength of the morphological evidence, my conclusion here is that chimpanzees and gorillas share a common ancestry separate from humans, and together form the sister-group to humans.

ACKNOWLEDGEMENTS

I am very grateful to Colin Patterson for the opportunity to present this paper and for his critical comments on the manuscript. I would also like to thank Charles Sibley for many good-natured exchanges on the relative merits of DNA and morphology, and I am grateful to Drs Terry Harrison, Lawrence Martin and Chris Stringer for their comments.

REFERENCES

Aiello, L. C. (1981). Locomotion in the Miocene Hominoidea. In *Aspects of Human Evolution*, ed. C. B. Stringer, pp. 63–97. London: Taylor & Francis.

Aiello, L. C. & Day, M. H. (1982). The evolution of locomotion in the early Hominidae. In *Progress in Anatomy*, ed. R. J. Harrison & V. Navaratnan, pp. 81–97. Cambridge University Press.

Andrews, P. (1985*a*). Family group systematics and evolution among catarrhine primates. In *Ancestors: the Hard Evidence*, ed. E. Delson, pp. 14–22. New York: Alan Liss.

Andrews, P. (1985*b*). Molecular evidence for catarrhine evolution. In *Major Topics in Primate and Human Evolution*, ed. B. Wood, L. Martin & P. Andrews, pp. 107–29. Cambridge University Press.

Andrews, P. & Cronin, J. (1982). The relationships of *Sivapithecus* and *Ramapithecus* and the evolution of the orang-utan. *Nature, London*, **297**, 541–6.

Andrews, P. & Groves, C. P. (1975). Gibbons and brachiation. In *Gibbon and Siamang*, vol. 4, ed. D. M. Rumbaugh, pp. 167–218. Basel: Karger.

Baba, M., Darga, L. & Goodman, M. (1980). Biochemical evidence on the phylogeny of the Anthropoidea. In *Evolutionary Biology of the New World Monkeys and Continental Drift*, ed. R. L. Ciochon & R. S. Corruccini, pp. 423–43. New York: Plenum Press.

Bishop, M. J. & Friday, A. E. (1985). Molecular sequences and hominoid phylogeny. In *Major Topics in Primate and Human Evolution*, ed. B. Wood, L. Martin & P. Andrews, pp. 150–6. Cambridge University Press.

de Boer, L. E. M. & Seuanez, H. N. (1982). The chromosomes of the orang utan and their relevance to the conservation of the species. In *The Orang Utan*, ed. L. E. M. de Boer, pp. 135–70. The Hague: Junk.

Bonde, N. (1984). Primitive features and ontogeny in phylogenetic reconstructions.

Videnskabelige Meddelelser fra Dansk Naturhistorisk Forening i Kjøbenhavn, **145**, 219–36.

Brown, W. M., Prager, E. M., Wang, A. & Wilson, A. C. (1982). Mitochondrial DNA sequences of primates: tempo and mode of evolution. *Journal of Molecular Evolution*, **18**, 225–39.

Bruce, E. J. & Ayala, F. J. (1979). Phylogenetic relationships between man and the apes: electrophoretic evidence. *Evolution*, **33**, 1040–56.

Cave, A. J. E. & Haynes, R. W. (1940). The paranasal sinuses of the anthropoid apes. *Journal of Anatomy*, **74**, 493–523.

Chang, L.-Y. E. & Slightom, J. L. (1984). Isolation and nucleotide sequence analysis of the B-type globin pseudogene from human, gorilla and chimpanzee. *Journal of Molecular Biology*, **180**, 767–84.

Corruccini, R., Baba, M., Goodman, M., Ciochon, R. & Cronin, J. (1980). Nonlinear macromolecular evolution and the molecular clock. *Evolution*, **34**, 1216–9.

Cross, J. F. & Martin, R. D. (1981). Calculation of gestation period and other reproductive parameters for primates. *Dodo*, **18**, 30–43.

Dean, M. C. & Wood, B. A. (1984). Phylogeny, neoteny and growth of the cranial base in hominoids. *Folia Primatologia*, **43**, 157–80.

Delson, E. & Andrews, P. (1975). Evolution and interrelationships of the catarrhine primates. In *Phylogeny of the Primates: an Interdisciplinary Approach*, ed. W. P. Luckett & F. S. Szalay, pp. 405–46. New York: Plenum Press.

Dene, H. T., Goodman, M. & Prychodko, W. (1976). Immunodiffusion evidence on the phylogeny of the primates. In *Molecular Anthropology*, ed. M. Goodman & R. E. Tashian, pp. 171–95. New York: Plenum Press.

Doolittle, R. F., Wooding, G. L., Lin, Y. & Riley, M. (1971). Hominoid evolution as judged by fibrinopeptide structures. *Journal of Molecular Evolution*, **1**, 74–83.

Farris, J. S. (1981). Distance data in phylogenetic analysis. In *Advances in Cladistics*, ed. V. A. Funk & D. R. Brooks, pp. 3–23. New York Botanical Garden.

Felsenstein, J. (1981). A likelihood approach to character weighting and what it tells us about parsimony and compatibility. *Biological Journal of the Linnean Society*, **16**, 183–96.

Felsenstein, J. (1983). Statistical inference of phylogenies. *Journal of the Royal Statistical Society*, **146**, 246–72.

Ferris, S. D., Sage, R. D., Huang, C. M., Nielson, J. T., Ritte, V. & Wilson, A. C. (1983). Flow of mtDNA across species boundaries. *Proceedings of the National Academy of Sciences USA*, **80**, 2290–4.

Ferris, S. D., Wilson, A. C. & Brown, W. M. (1981*a*). Evolutionary tree for apes and humans based on cleavage maps of mtDNA. *Proceedings of the National Academy of Sciences USA*, **78**, 2432–6.

Ferris, S. D., Brown, W. M., Davidson, W. S. & Wilson, A. C. (1981*b*). Extensive polymorphism in the mtDNA of apes. *Proceedings of the National Academy of Sciences USA*, **78**, 6319–23.

Frelin, C. & Vuilleumier, F. (1979). Biochemical methods and reasoning in systematics. *Zeitschrift für Zoologische Systematik und Evolutionsforschung*, **17**, 1–10.

Friday, A. E. (1981). Hominoid evolution: the nature of the biochemical evidence. In

Aspects of Human Evolution, ed, C. B. Stringer, pp. 1–23. London: Taylor & Francis.

Giles, E. (1956). Cranial allometry in the great apes. *Human Biology*, **28**, 43–58.

Goodman, M. (1963). Man's place in the phylogeny of the primates as reflected in serum proteins. In *Classification and Human Evolution*, ed. S. L. Washburn, pp. 204–34. Chicago: Aldine Press.

Goodman, M. (1973). The chronicle of primate phylogeny contained in proteins. *Symposium of the Zoological Society of London*, **33**, 339–75.

Goodman, M. (1974). Biochemical evidence on hominid phylogeny. *Annual Review of Anthropology*, **3**, 203–26.

Goodman, M. (1975). Protein sequence and immunological specificity: their role in phylogenetic studies of primates. In *Phylogeny of the Primates: an Interdisciplinary Approach*, ed. W. P. Luckett & F. S. Szalay, pp. 219–48. New York: Plenum Press.

Goodman, M. (1976). Towards a genealogical description of the primates. In *Molecular Anthropology*, ed. M. Goodman & R. E. Tashian, pp. 321–53. New York: Plenum Press.

Goodman, M. (1981). Decoding the pattern of protein evolution. *Progress in Biophysical and Molecular Biology*, **38**, 105–64.

Goodman, M., Baba, M. & Darga, L. (1983*a*). The bearing of molecular data on the cladogenesis and times of divergence of hominoid lineages. In *New Interpretations of Ape and Human Evolution*, ed. R. L. Ciochon & R. S. Corruccini, pp. 67–86. New York: Plenum Press.

Goodman, M., Braunitzer, G., Stangl, A. & Schank, B. (1983*b*). Evidence on human origins from haemoglobins of African apes. *Nature, London*, **303**, 546–8.

Goodman, M., Koop, B. F., Czelusniak, J., Weiss, M. L. & Slightom, J. L. (1984). The η-globin gene: its long evolutionary history in the β-globin gene family of mammals. *Journal of Molecular Biology*, **180**, 803–23.

Harris, S., Barrie, P. A., Weiss, M. L. & Jeffreys, A. (1984). The primate *YB1* gene. *Journal of Molecular Biology*, **180**, 785–801.

Harrison, T. (1982). Small-bodied apes from the Miocene of East Africa. PhD thesis, University of London.

Hasegawa, M., Yano, T. & Kishino, H. (1984). A new molecular clock of mtDNA and the evolution of hominoids. *Proceedings of the Japan Academy*, **60**, 95–8.

Hasegawa, M., Kishino, H. & Yano, T. (1985). Dating of human–ape splitting by a molecular clock of mtDNA. *Journal of Molecular Evolution*, **22**, 160–74.

Hoyer, B. H., van de Velde, N. W., Goodman, M. & Roberts, R. B. (1972). Examination of hominoid evolution by DNA sequence homology. *Journal of Human Evolution*, **1**, 645–9.

Kluge, A. G. (1983). Cladistics and the classification of the great apes. In *New Interpretations of Ape and Human Ancestry*, ed. R. L. Ciochon & R. S. Corruccini, pp. 151–77. New York: Plenum Press.

Mai, L. L. (1983). A model of chromosome evolution and its bearing on cladogenesis in the Hominoidea. In *New Interpretations of Ape and Human Ancestry*, ed. R. L. Ciochon & R. S. Corruccini, pp. 87–114. New York: Plenum Press.

Marks, J. (1983). Hominoid cytogenetics and evolution. *Yearbook of Physical Anthropology*, **26**, 131–59.

Martin, L. B. (1983). The relationships of the later Miocene Hominoidea. PhD thesis, University of London.

Martin, L. B. (1985*a*). The significance of enamel thickness in hominoid evolution. *Nature, London*, **314,** 260–3.

Martin, L. B. (1985*b*). Relationships among extant and extinct great apes and humans. In *Major Topics in Primate and Human Evolution*, ed. B. Wood, L. Martin & P. Andrews, pp. 161–87. Cambridge University Press.

Martin, R. D. & MacLarnon, A. (1985). Gestation period, neonatal size and maternal investment in placental mammals. *Nature, London*, **313,** 220–3.

Miller, D. A. (1977). Evolution of primate chromosomes. *Science*, **198,** 1116–24.

Miyata, T., Hayashida, H., Kikuno, R., Hasegawa, M., Kobayashi, M. & Koike, K. (1982). Molecular clock of silent substitution: at least six-fold preponderance of silent changes in mitochondrial genes over those in nuclear genes. *Journal of Molecular Evolution*, **19,** 18–35.

Perutz, M. F. (1983). Species adaptation in a protein molecule. *Molecular Biology and Evolution*, **1,** 1–28.

Read, D. W. (1975). Primate phylogeny, neutral mutations and 'molecular clocks'. *Systematic Zoology*, **24,** 209–21.

Read, D. W. & Lestrel, P. E. (1970). Hominid phylogeny and immunology: a critical appraisal. *Science*, **168,** 578–80.

Romero-Herrera, A. E., Lehmann, H., Castillo, O., Joysey, K. A. & Friday, A. E. (1976). Myoglobin of the orang utan as a phylogenetic enigma. *Nature, London*, **261,** 162–4.

Romero-Herrera, A. E., Lehmann, H., Joysey, K. A. & Friday, A. E. (1978). On the evolution of myoglobin. *Philosophical Transactions of the Royal Society, B*, **283,** 61–163.

Ruvolo, M. (1983). Genetic evolution in the African guenon monkeys (Primates, Cercopithecinae). PhD thesis, Harvard University.

Ruvolo, M. & Pilbeam, D. R. (1985). Hominoid evolution: molecular and palaeontological pattern. In *Major Topics in Primate and Human Evolution*, ed. B. Wood, L. Martin & P. Andrews, pp. 157–60. Cambridge University Press.

Sarich, V. M. & Cronin, J. E. (1976). Molecular systematics of the primates. In *Molecular Anthropology*, ed. M. Goodman & R. E. Tashian, pp. 141–70. New York: Plenum Press.

Sarich, V. M. & Wilson, A. C. (1967). Immunological time-scale for hominid evolution. *Science*, **158,** 1200–3.

Schwartz, J. H. (1984). On the evolutionary relationships of humans and orang utans. *Nature, London*, **308,** 501–5.

Sibley, C. G. & Ahlquist, J. E. (1984). The phylogeny of the hominoid primates, as indicated by DNA–DNA hybridization. *Journal of Molecular Evolution*, **20,** 2–15.

Socha, W. W. & Ruffié, J. (1983). *Blood Groups of Primates: Theory, Practice, Evolutionary Meaning*. New York: Alan Liss.

Spolsky, C. & Uzzell, T. (1984). Natural interspecies transfer of mtDNA in amphibians. *Proceedings of the National Academy of Sciences USA*, **81,** 5802–5.

Stanyon, R. & Chiarelli, B. (1982). Phylogeny of the Hominoidea: the chromosomal evidence. *Journal of Human Evolution*, **11,** 493–504.

Tashian, R. E., Goodman, M., Ferrell, R. E. & Tanis, R. J. (1976). Evolution of carbonic anhydrase in primates and other mammals. In *Molecular Anthropology*, ed. M. Goodman & R. E. Tashian, pp. 301–19. New York: Plenum Press.

Templeton, A. R. (1983). Phylogenetic inference from restriction endonuclease cleavage site maps with particular reference to the evolution of humans and the apes. *Evolution*, **37,** 221–44.

Tuttle, R. H. (1967). Knuckle-walking and the evolution of hominoid hands. *American Journal of Physical Anthropology*, **26,** 171–206.

Tuttle, R. H. (1969). Quantitative and functional studies on the hands of the Anthropoidea. *Journal of Morphology*, **128,** 309–64.

Ward, S. C. & Pilbeam, D. R. (1983). Maxillofacial morphology of Miocene hominoids from Africa and Indo-Pakistan. In *New Interpretations of Ape and Human Ancestry*, ed. R. L. Ciochon & R. S. Corruccini, pp. 211–38. New York: Plenum Press.

White, M. J. D. (1973). *Animal Cytology and Evolution*. Cambridge University Press.

Wu, C.-I. & Li, W.-H. (1985). Evidence for higher rates of nucleotide substitution in rodents than in man. *Proceedings of the National Academy of Sciences USA*, **82,** 1741–5.

Yunis, J. J. & Prakash, O. (1982). The origin of man: a chromosomal pictorial legacy. *Science*, **215,** 1525–30.

3 Molecular and morphological analysis of high-level mammalian interrelationships

Malcolm C. McKenna

Introduction

Crick (1958) suggested that '... before long we shall have a subject which might be called "protein taxonomy" ...'. He was correct. In mammals, analysis of the pattern of evolution of amino acid sequences in several polymorphic protein molecules has now been carried to the point where an assessment of congruence can be made with regard to evolutionary pattern as seen by anatomists and palaeontologists. For current examples of molecular results, see Goodman, Czelusniak & Beeber (1985; and literature cited therein).

Do cladograms depicting hypotheses of successive substitution of amino acids in the molecular history of mammals agree with notions of evolutionary pattern of evolving populations at the morphological level? If not, why not? In cases of disagreement, are the chemists wrong? The morphologists? Both? Or are the two approaches headed for some sort of eclectic merger?

This paper examines afresh the amino acid sequence data from two biologically interesting molecules occurring in vertebrates: myoglobin and the alpha crystallin A chain. Myoglobin is a protein that stores oxygen in muscle (Romero-Herrera *et al.*, 1978; Fermi & Perutz, 1981). The alpha crystallin A chain (de Jong & Goodman, 1982) is a slowly evolving lens protein of the vertebrate eye. Maximally parsimonious character phylogenies embedded in evolutionary phylogenies are sought and then compared. However, there is no guarantee that nature has followed exactly the track laid out (Hull, 1979).

The analyses reported here were made mainly in 1978 and 1980 without the aid of a computer, but a few adjustments have been made from time to time since then as new vertebrates were sequenced or as errors became evident (e.g. those in the chicken myoglobin sequence; Prass, Berkley & Romero-Herrera, 1983). My results are closely similar to the various cladograms generated from the same and other data by Morris Goodman and his associates (e.g. Goodman *et al.*, 1985), but differ in that I allow trichotomies or polytomies in my cladograms rather than demand that they be fully dichotomous. Frequently, the data under analysis will not resolve a trichotomous or polytomous node into one of the many dichotomous schemes compatible with it (e.g. Nelson & Platnick, 1981, Figs 3.2–3.11, 3.20). Goodman believes that cladograms should be dichotomous (i.e. fully resolved) and, when the data under analysis do not resolve polytomies, he draws on other evidence to resolve them, producing dichotomous cladograms including zero-distance limbs (e.g. de Jong & Goodman, 1982, Fig. 5). Also, rather than a mere enumeration of how many substitutions are postulated at a given locus, the hypothetical substitutions themselves are listed here (Appendix 1) for each node of the cladogram in order to increase accessibility and vulnerability. One cannot use Goodman's cladograms as a starting point for further work; instead, one must use the data base. Nevertheless, my cladograms can be regarded as an independent check of his excellent work.

In mammals, myoglobin possesses a sequence of 153 amino acid residues and the alpha crystallin A chain has 173. Among the organisms studied, of course, not all of these sites have been subjected to amino acid substitutions. Both proteins can be collected in abundance from large vertebrates, and can be had in useful amounts by pooling samples drawn from several individuals of small vertebrates. At present, because of the expense, few taxa have been analysed more than once, so that possible molecular polymorphism within a species is ignored and errors based on extrapolation from a single individual (which might be atypical in the instance sampled) could reside unchallenged in the compiled tables of sequence data for species. Moreover, in some taxonomically diverse supraspecific groups analysed on the basis of one species, that species may not be representative. Such double synecdoche must be tolerated until additional diversity of both individuals and species can be sampled in order to establish more clearly the most parsimonious historical sequence of hypothesized changes. For the closely related molecule haemoglobin, more than 200 variations in the amino acid sequence are known in humans (de Jong & Rydén, 1981), such as the substitution from glutamic acid to valine at position 6 in the beta chain (Fermi & Perutz, 1981, p. 74) that produces sickle-cell anaemia and resistance to malaria. Some of these even involve addition of as many as 11 residues at the C (downstream) terminus of beta chains. By analogy, if as much attention were directed to human and other myoglobins as has been lavished on human haemoglobin, additional variation would surely appear. It is also possible that some sequences are incorrectly determined, as was originally the case with the chicken (Prass *et al.*, 1983).

I have preferred to analyse the amino acid sequence data obtained by others from myoglobin and the alpha crystallin A chain rather than to rely on studies of serology of transferrins or on the few available studies of DNA hybridization, because at least with data from reasonably small and relatively inexpensively sequenced molecules one can confidently identify the basis of similarity at the amino acid residue level and often even at the codon level. Given sequences, one need not depend on an approach in which one knows only degrees of reactivity or temperatures at which complex chains of unknown sequence dissociate, a dubious mixture of phenetic and cladistic methodology (Carpenter, 1985).

The results of cladistic analysis of sequence diversity in myoglobin and the alpha crystallin A chain yield essentially a single major scheme of genealogical relationship, but the two resulting cladograms are not identical in geometry. Similar results have been obtained by Goodman *et al.* (1985) by combining even more data from cytochrome *c*, fibrinopeptides A and B, and alpha and beta haemoglobin in addition to myoglobin and the alpha crystallin A chain. As in all work, a tolerable number of minor infelicities remain to be sorted out and eliminated or explained, because at certain taxonomic levels different molecules still give divergent results in some cases.

It may be wondered why I have not used a branch-swapping computer program (Moore, 1975). My answer is that, although minor errors may persist because 'manual' methods are not as thorough as a branch-swapping program would be, when cladograms reach a certain level of complexity – far less than that of those developed here – branch-swapping programs used to construct them must themselves become arbitrary in order to avoid enormous running time and consequent expense. For example, 150 hours of CPU time on a Prime model 750 computer were required in a phylogenetic study of 5 proteins in just 11 mammal species studied by Penny & Hendy (1985*b*). However, processing time must be arbitrarily limited, as with computer chess algorithms. For 22 species, the number of rooted trees would be about 10^{29} (Felsenstein, 1978), a prohibitive number if every permutation and combination is to be considered, because the problem is what in computer science is called 'nondeterministic polynomial complete' (NP-complete). However, I have analysed more than 50 species with regard to both the myoglobin molecule and the alpha crystallin A chain. Of course, I have consciously ignored all but a few 'obvious' possibilities. This is because a human investigator can cheerfully run the risk attendant on not studying the possibility that fruit bats are descended from whales, but it is difficult to train a computer to do so without running the risk that a group of organisms like pangolins will *not* be compared with carnivores. At present, a human investigator can, in principle, abandon a local parsimony in favour of a more general one much earlier than a computer can. Recently, a similar conclusion was reached rigorously by Graham & Foulds (1982). Finally, I have not tried to minimize reversals during the construction of the cladograms presented here.

After discussing the molecular results, I then compare relationships hypothesized on the basis of molecular information with some schemes of relationship based upon morphological studies, notably those developed by palaeontologists working with a fossil record that ranges in quality from excellent to mediocre.

To make a long story short, the results are that molecular cladistic analyses of myoglobin and the alpha crystallin A chain yield results which, while they may differ in detail, are generally compatible with current palaeontological and morphological work but differ widely from some of the older ideas embalmed in textbooks. It should also be obvious that molecular studies can suffer from exactly the same ills that beset comparative anatomical ones: a touchstone has not been found that will test phylogeny unequivocally.

Myoglobin

A cladogram (Figs 3.1 to 3.6, solid lines) based on myoglobin amino acid sequences taken from 62 species of tetrapods ranging from turtles to mammals was constructed from data kindly supplied by Morris Goodman and Alejo Romero-Herrera (see also Romero-Herrera *et al.*, 1978). Most of the assignment of substitutions to various nodes can be done by inspection or with a word processor as a simple aid in comparing sequences or in searching for parallelisms. All postulated amino acid substitutions were checked at the codon (mRNA) level to determine the minimum number of 'hits' (nucleotide replacements = NRs) required to transform them. Table 3.1 lists a hypothetical sequence of amino acid residues for the 153 loci present in what I calculate to be the plesiomorphous condition for myoglobin in Mammalia. The plesiomorphous condition for each mammalian locus has been determined by outgroup comparison with myoglobin sequence data from eight species of non-mammalian vertebrates, including three species of fishes not discussed in this paper. Only the more derived part of the more inclusive cladogram is used for present purposes.

Mammalia

As expected, the living mammalian representatives of theropsid tetrapods are confirmed as a monophyletic group (Fig. 3.1: node 11) sharing 12 amino acid substitutions in their myoglobin, derived from the hypothetical plesiomorphous sequence (Table 3.1). Changes at loci 113 (lysine to histidine), 116 (alanine to glutamine), and 117 (glutamic acid to serine) represent the end result of double 'hits'. Thus, there is actually a minimum number of 15 shared nucleotide replacements. Moreover, although not the subject of this paper, the sauropsid tetrapods (*sensu* Goodrich, 1916) are also confirmed as monophyletic (Fig. 3.1: node 2); however, the internal branching pattern of

Table 3.1. *Myoglobin amino acid sequence hypothesized to be primitive for tetrapods (active loci underlined)*

−1 ?	23 gly	46 phe	69 leu	92 ser	115 ile	138 phe
1 gly	24 his	47 lys	70 thr	93 his	116 ala	139 arg
2 leu	25 gly	48 his	71 ala	94 ala	117 glu	140 asn
3 ser	26 gln	49 leu	72 leu	95 thr	118 lys	141 asp
4 asp	27 glu	50 lys	73 gly	96 lys	119 his	142 met
5 gly	28 val	51 thr	74 asn	97 his	120 pro	143 ala
6 glu	29 leu	52 glu	75 ile	98 lys	121 ala	144 ala
7 trp	30 ile	53 asp	76 leu	99 ile	122 asp	145 lys
8 gln	31 arg	54 glu	77 lys	100 pro	123 phe	146 tyr
9 gln	32 leu	55 met	78 lys	101 val	124 gly	147 lys
10 val	33 phe	56 lys	79 lys	102 lys	125 ala	148 glu
11 leu	34 lys	57 ala	80 gly	103 tyr	126 asp	149 phe
12 asn	35 gly	58 ser	81 his	104 leu	127 ala	150 gly
13 val	36 his	59 glu	82 his	105 glu	128 gln	151 phe
14 trp	37 pro	60 asp	83 glu	106 phe	129 ala	152 gln
15 gly	38 glu	61 met	84 ala	107 ile	130 ala	153 gly
16 lys	39 thr	62 lys	85 glu	108 ser	131 met	
17 val	40 gln	63 lys	86 leu	109 glu	132 lys	
18 glu	41 glu	64 his	87 lys	110 val	133 lys	
19 ala	42 arg	65 gly	88 pro	111 ile	134 ala	
20 asp	43 phe	66 ala	89 leu	112 ile	135 leu	
21 leu	44 ala	67 thr	90 ala	113 lys	136 glu	
22 pro	45 lys	68 val	91 gln	114 val	137 leu	

the sauropsids obtained here on the basis of myoglobin is not what was expected (e.g. Benton, 1985) and is contradicted by the alpha crystallin A chain cladogram (position of the alligator with respect to the two lizards). Nevertheless, Gardiner's (1982) alliance of birds and mammals in the taxon Haemothermia is firmly contradicted (cf. Bishop & Friday, this volume, pp. 124–37).

Monotremes and therians

Within the Mammalia, myoglobin of the living monotremes of Australia and New Guinea singles out these animals as the first major branch, again not an unexpected conclusion. However, in a paper published as recently as 1985, Goodman *et al.* (1985) achieve a very curious result: in figures 1 and 3 of their paper, they place the monotremes within the Eutheria after the phylogenetic departure of the elephants and edentates, thus outdoing even Gregory's (1947) 'palimpsest theory', the belief that monotremes are specialized marsupials. Elsewhere in the same paper, Goodman and his colleagues get more acceptable results. In the cladogram presented here, in contrast to therians, the platypus (*Ornithorhynchus*) and the echidna (*Tachyglossus*) share 10 amino acid substitutions in their myoglobin, 2 of which are double 'hits' (Fig. 3.1: node 12). That the two major kinds of monotremes have had long independent histories is suggested by three autapomorphies (one a double 'hit') in myoglobin of the platypus and seven in the echidna.

The myoglobin of living therian mammals (placentals = eutherians, plus marsupials) possesses only five synapomorphies separating them from the monotremes and still more primitive living vertebrates. It is interesting that fewer substitutions in the inferred ancestral myoglobin sequence separate the therians from the monotremes than separate 'higher' artiodactyls from pigs (*Sus scrofa*), or Anthropoidea from lemurs and lorises. Thus, Gregory (1947) was fairly close to the truth when he allied monotremes closely with the

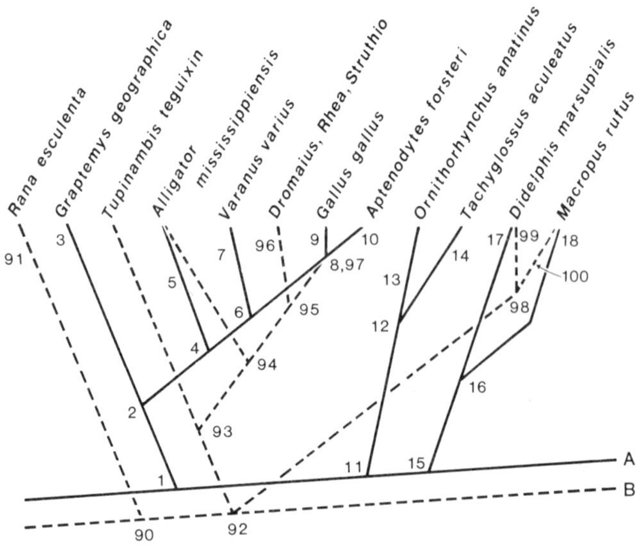

Fig. 3.1 Superposed cladograms of hypothesized phylogenetic changes in tetrapod myoglobin (solid lines) and the alpha crystallin A chain (dashed lines). Part one: amphibians to marsupials. A and B connect to A and B of Fig. 3.2. See Appendix 1 for explanation of the numbers at the nodes. Common names of genera are: *Rana*, frog; *Graptemys*, turtle; *Tupinambis*, tegu lizard; *Varanus*, monitor lizard; *Dromaius*, emu; *Struthio*, ostrich; *Gallus*, chicken; *Aptenodytes*, penguin; *Ornithorhynchus*, platypus; *Tachyglossus*, echidna; *Didelphis*, American opossum; *Macropus*, kangaroo.

ancestors of marsupials, but was wrong when he formalized the paraphyletic taxon Marsupionta to contain monotremes and marsupials.

Next, the Theria (Fig. 3.1: node 15) can be divided on the basis of myoglobin sequence data into the traditionally accepted divisions: Metatheria (order Marsupialia of most authors) and Eutheria (placental mammals). The American opossum (*Didelphis*) and the Australian red kangaroo (*Macropus rufus*) share four synapomorphous substitutions, but they have diverged, with development of a large number of autapomorphies in the myoglobin of each clade. This might be expected in groups with such a long history of geographic isolation from one another. Eutherian myoglobin shares even fewer synapomorphous substitutions (Fig. 3.2: node 19): locus 74 to glycine (a double 'hit'), 103 to tyrosine (a back mutation), and 142 to isoleucine. This gives a total of four substitutions, the same number holding together the two marsupials tested.

It is of interest that the metatherians have generally been placed in a single order, whereas the eutherians are usually awarded more than 30.

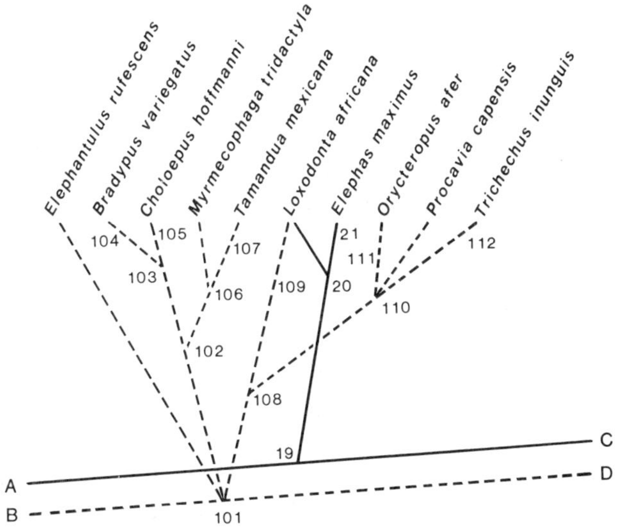

Fig. 3.2 Superposed cladograms of hypothesized phylogenetic changes in tetrapod myoglobin (solid lines) and the alpha crystallin A chain (dashed lines). Part two: *Elephantulus* to *Trichechus*. A and B connect to A and B of Fig. 3.1. C and D connect to C and D of Fig. 3.3. See Appendix 1 for explanation of the numbers at the nodes. Common names of genera are: *Elephantulus*, elephant shrew; *Bradypus*, *Choloepus*, sloths; *Myrmecophaga*, *Tamandua*, anteaters; *Loxodonta*, *Elephas*, elephants; *Orycteropus*, aardvark; *Procavia*, hyrax; *Trichechus*, manatee. *Orycteropus* (myoglobin cladogram) is repeated on Fig. 3.5.

Eutherian divergences

Within the living Eutheria, myoglobin has been sequenced in seven major groups: Proboscidea, Carnivora, Cetacea, Perissodactyla, Artiodactyla, Primates, and a poorly resolved assemblage that includes Scandentia (*Tupaia*, treeshrews), Lagomorpha (*Ochotona*, pika, and *Oryctolagus*, rabbit), Chiroptera (only the fruit bat *Rousettus* so far), Tubulidentata (*Orycteropus*, aardvark), and Insectivora (only the hedgehog *Erinaceus* so far). Unfortunately, several major still-living eutherian groups that should be included in an adequate sample can not be. We lack myoglobin sequence data for Edentata (sloths, anteaters, armadillos), Hyracoidea (hyrax, 'coney', dassie, etc.), Sirenia (dugongs, manatees), Pholidota (scaly anteater), Dermoptera ('flying lemur') and Macroscelidea (elephant shrews). Among the major groups in which at least one member has had its myoglobin sequenced, it is important to study additional subtaxa, especially in the Macroscelidea, Carnivora (e.g. in bears, procyonids, both kinds of panda, felids, viverrids, hyaenids), Perissodactyla (tapirs, rhinos), and Artiodactyla (camels, llamas, hippos, peccaries, tragulids, antelopes, antilocaprids, giraffes). Of all these, probably the macroscelideans (especially *Rhynchocyon*) and tragulids would be the most interesting and useful.

The first therian branch represented among living mammals in which myoglobin has been sampled is the order Proboscidea (elephants), one of the orders of Simpson's (1945) superorder Paenungulata. The other living paenungulates, Hyracoidea and Sirenia, have not had their myoglobin sequenced yet, nor have the edentates, which on the basis of the alpha crystallin A chain (below) might be expected to depart from the main eutherian stem at about the same time as, or before, the elephants did. Tubulidentata (aardvarks), although sometimes allied with living paenungulates, are not grouped with them on the basis of myoglobin. However, the alpha crystallin A chain does suggest a similarity to the Proboscidea.

Both living genera of elephants have been studied, but in the myoglobin molecule they differ from one another at only one locus. The Indian elephant, *Elephas*, has undergone a substitution at locus 129 to glycine since splitting from the line leading to the African elephant, *Loxodonta*. In contrast, myoglobin in the two extant genera of elephants shares 14 synapomorphous substitutions (Fig. 3.2: node 20), suggesting a very long period of genetic isolation of Proboscidea from other mammals studied thus far. Needless to say, results from edentates, pangolins, hyraxes and sirenians would be most welcome.

After the phylogenetic departure of the elephants (and probably the other paenungulates as well), myoglobin of the rest of the living mammals under review primitively shares (Fig. 3.3: node 22) only three substitutions: at loci 21, 22 and 66. Locus 21 underwent a back mutation, locus 66 took a double 'hit', and all three of these loci are subject to parallelism elsewhere. I do not attach much weight to these synapomorphies, but they can be tested easily by adding sequence data from myoglobin of edentates, elephant shrews, hyraxes and sirenians to the cladogram, as has been done in studies of the evolution of lens alpha crystallin A chain.

A division occurs between myoglobin of Carnivora on the one hand (Fig. 3.3: node 23) and Cetacea, Perissodactyla, Artiodactyla, Primates, and the aardvark–treeshrew–rabbit–bat–hedgehog collocation on the other (Fig. 3.4: node 31). The Carnivora are monophyletic because of synapomorphous substitutions at loci 19, 35 and 57; within the Carnivora, the canoids (dogs; Fig. 3.3: node 24) and arctoids (bear-like animals, including pinnipeds: seals, walruses and sealions; Fig. 3.3: node 26) are separate, but differentiation within the arctoid carnivorans appears to have occurred earlier than in the

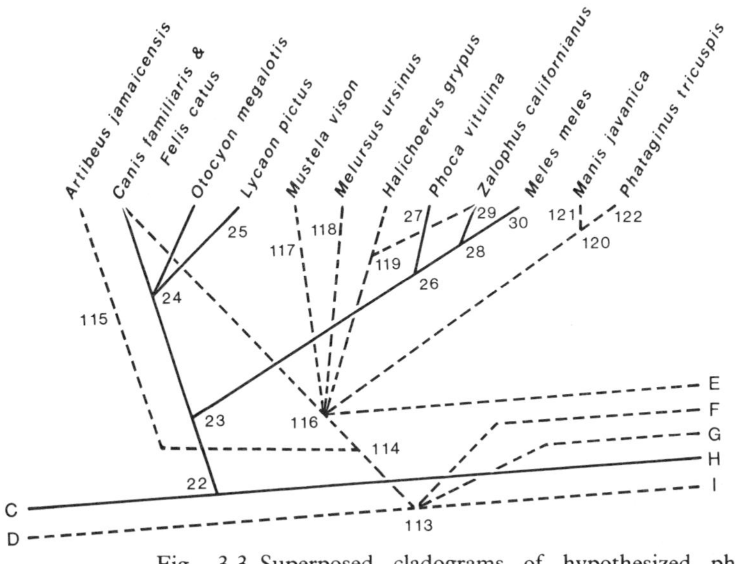

Fig. 3.3 Superposed cladograms of hypothesized phylogenetic changes in tetrapod myoglobin (solid lines) and the alpha crystallin A chain (dashed lines). Part three: *Artibeus* to *Phataginus*. C and D connect to C and D of Fig. 3.2. E to I connect to E to I of Fig. 3.4. See Appendix 1 for explanation of the numbers at the nodes. [One NR can be saved if the sloth bear (*Melursus*) and the two pangolins (*Manis* and *Phataginus*) are assumed to share 74 tyr= as a synapomorphy. See nodes 118 and 120, Appendix 1.] Common names of genera are: *Artibeus*, American fruit bat (a microchiropteran bat); *Canis*, dog; *Felis*, cat; *Otocyon*, bat-eared fox; *Lycaon*, hunting dog; *Mustela*, weasel; *Melursus*, sloth bear; *Halichoerus*, *Phoca*, seals; *Zalophus*, sealion; *Meles*, badger; *Manis*, *Phataginus*, pangolins.

canoids sampled, because myoglobin of each arctoid line has many autapomorphous substitutions and because there are few synapomorphous ones. If myoglobin were to be sequenced in additional living arctoids, this would test the few synapomorphous substitutions that appear to join those few from which sequences are known.

After the departure of the carnivorans, the remaining mammals are hypothesized to share only three myoglobin substitutions primitively (Fig. 3.4: node 31), one of them at the apparently easily mutated locus 21, where a substitution to valine took place.

At this point in the myoglobin cladogram (Fig. 3.4: node 32), a strange collocation of Perissodactyla (horses and allies, node 51) and Cetacea (whales, dolphins, node 33) breaks away, characterized by apparently synapomorphous myoglobin substitutions at locus 53 (to alanine), 101 (to isoleucine, a back mutation), and 116 (to histidine). The line leading to horses and zebras departed early from one leading to the cetaceans, developing six autapomorphous substitutions (one a double 'hit' and another a back mutation; Fig. 3.4:

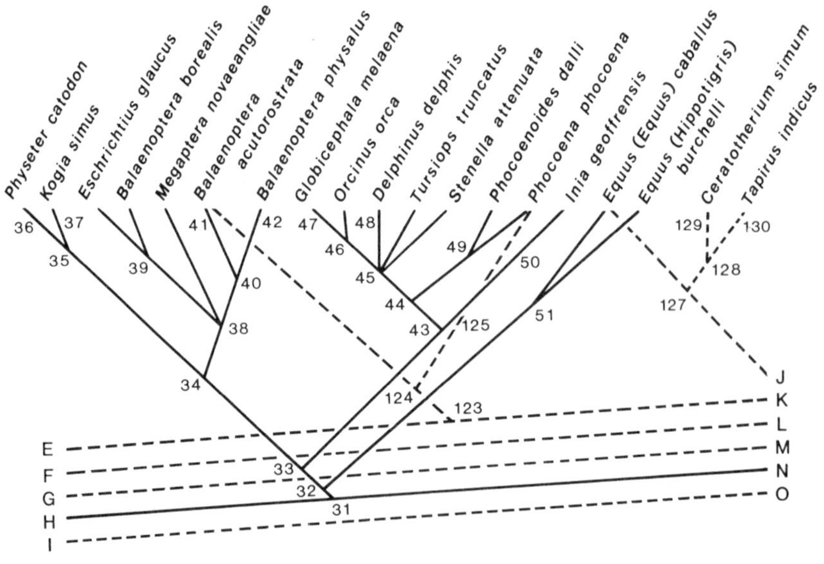

Fig. 3.4 Superposed cladograms of hypothesized phylogenetic changes in tetrapod myoglobin (solid lines) and the alpha crystallin A chain (dashed lines). Part four: *Physeter* to *Tapirus*. E to I connect to E to I of Fig. 3.3. J to O connect to J to O of Fig. 3.5. See Appendix 1 for explanation of the numbers at the nodes. Common names of genera are: *Physeter*, *Kogia*, sperm whales; *Eschrichtius*, grey whale; *Balaenoptera*, *Megaptera*, rorquals; *Globicephala*, pilot whale; *Orcinus*, killer whale; *Delphinus*, *Tursiops*, *Stenella*, dolphins; *Phocoenoides*, *Phocoena*, porpoises; *Inia*, Amazon porpoise; *Equus*, horse; *Hippotigris*, zebra; *Ceratotherium*, rhinoceros; *Tapirus*, tapir.

node 51). In serological studies (J. Shoshani, M. Goodman & W. Prychodko, unpublished data) and in the alpha crystallin A chain cladogram discussed below and shown by broken lines in Fig. 3.4, the perissodactyls shift from being the sister-group of the cetaceans to be the sister-group of artiodactyls. One of these solutions must be incorrect, but the error is minor compared to the results shown in figures 1 and 3 of Goodman *et al.* (1985), wherein camel, llama and a whale are depicted as the sister-group of horses, asses, zebras, sheep and oxen. Such a scheme would make Perissodactyla and Cetacea into subdivisions of the Artiodactyla (or make artiodactyls paraphyletic). Although these three ungulate orders are related, their anatomy – notably their dentition and foot structure – prevents their being shuffled in such a manner. One should follow a computer only up to the *edge* of an intellectual cliff.

Myoglobin of living cetaceans developed a set of five synapomorphous substitutions (Fig. 3.4: node 33): at loci 27 (to aspartic acid), 109 (to glutamic acid, a back mutation), 118 (to arginine), 121 (to alanine, another back mutation), and 140 (to lysine). Within the Cetacea, a rich array of 15 species has been sampled; an interesting section of the myoglobin cladogram results. The genera of various extant cetacean families hold together in the way predicted by morphological studies, but it appears that the monophyletic mysticetes (baleen whales, node 38) originated from within a paraphyletic group of 'odontocetes' (plesiomorphously toothed whales), only some of which (*Physeter*, sperm whale; *Kogia*, pygmy sperm whale: family Physeteridae) are the sister-group of the mysticetes on the basis of synapomorphous substitutions in myoglobin (Fig. 3.4: node 34) at loci 1 (to valine), 15 (to alanine), and 28 (to isoleucine). On the basis of the postulated evolutionary history of myoglobin, this combined group of mysticetes and sperm whales is the sister-group of the dolphins (node 43). If these relationships stand up to further sequence sampling and analysis, either the 'Odontoceti' should be abandoned or the Mysticeti should be merged with them as a monophyletic subdivision. It would also appear that the species of *Balaenoptera* need re-examination because this genus appears to be polyphyletic (see Fig. 3.4: nodes 38, 39 and 40). Such a new way of looking at the classification of cetaceans is not a conflict between morphology and molecular biology.

The remainder of the living eutherians tested are characterized very weakly (Fig. 3.5: node 52) by a pair of synapomorphous substitutions in the myoglobin molecule. At locus 109, glutamic acid is substituted for asparagine and at locus 132 serine is substituted for asparagine. The mutation to glutamic acid at locus 109 is a back mutation to the condition pertaining in monotremes, a state assumed to be plesiomorphous for Mammalia.

The first branch within this remaining mammalian group comprises the Artiodactyla. From the earliest known Eocene examples to the present, these animals are easily recognized on morphological grounds (e.g. the unique astragalo-calcaneal complex: Schaeffer, 1947). However, with regard to myoglobin, they are held together only by a single synapomorphous back mutation (Fig. 3.5: node 53) at locus 142 to methionine, this time to the amino acid pertaining not only in monotremes but also in marsupials. Again, this is presumably the plesiomorphous mammalian condition. The extant artiodactyls tested display a strange pattern of myoglobin substitutions, whose most parsimonious geometry closely fits morphologically based conclusions but should give enthusiasts for molecular clocks something to think about. The fossil record proves that *Sus* (pig) and the euartiodactyls ('higher artiodactyls') split apart at some point prior to the Oligocene, at least 38 MyrBP, and they obviously have had equally long histories since then. However, only 2 myoglobin substitutions (113 to glutamine and 115 to isoleucine) characterize *Sus* (Fig. 3.5: node 54), whereas 22 substitutions involving 24 NRs characterize

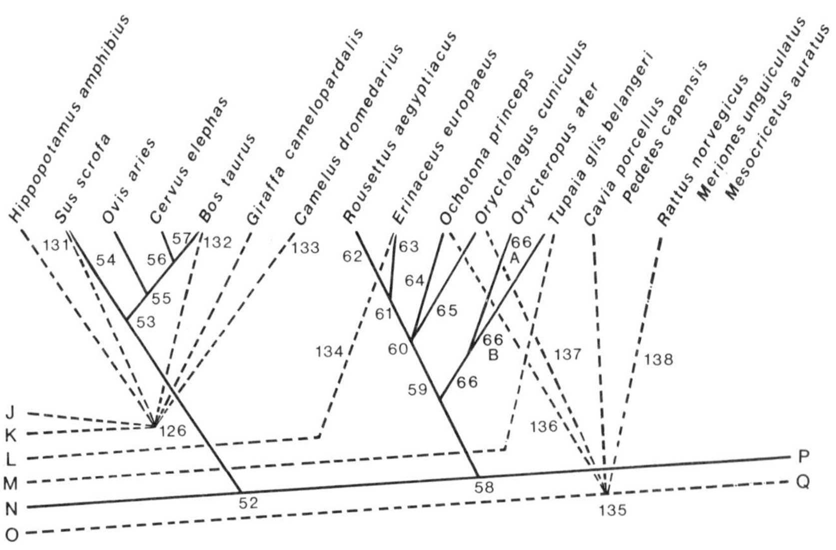

Fig. 3.5 Superposed cladograms of hypothesized phylogenetic changes in tetrapod myoglobin (solid lines) and the alpha crystallin A chain (dashed lines). Part five: *Hippopotamus* to *Mesocricetus*. J to O connect to J to O of Fig. 3.4. P and Q connect to P and Q of Fig. 3.6. See Appendix 1 for explanation of the numbers at the nodes. [*Orycteropus* (aardvark, order Tubulidentata) also appears in Fig. 3.2, where analysis of the alpha crystallin A chain places it as a sister-group of *Procavia* and *Trichechus*. See node 66 in Appendix 1 for details.] Common names of genera are: *Sus*, pig; *Ovis*, sheep; *Cervus*, deer; *Bos*, oxen; *Giraffa*, giraffe; *Camelus*, camel; *Rousettus*, fruit bat (a megachiropteran bat); *Erinaceus*, hedgehog; *Ochotona*, pika; *Oryctolagus*, rabbit; *Tupaia*, treeshrew; *Cavia*, guinea pig; *Pedetes*, springhaas; *Rattus*, rat; *Meriones*, gerbil; *Mesocricetus*, hamster.

Bos (cattle; Fig. 3.5: nodes 55 and 56). Seven of these 'hits' (Fig. 3.5: node 56) are autapomorphous and occurred after the separation of *Ovis* and *Bos*, in the late Cenozoic, yet *Ovis* (sheep) has no indicated myoglobin autapomorphies. Moreover, it is instructive that locus 101 in the myoglobin of *Ovis* is valine, not isoleucine (Vötsch & Anderer, 1975). If this is a correction of earlier work and not just the discovery of polymorphism in *Ovis*, and is taken as a locally parsimonious symplesiomorphy, then the cladogram saves one NR. However, reversal to valine from isoleucine at this locus in *Ovis*, after an earlier synapomorphous mutation to isoleucine in the common ancestor of *Ovis*, *Bos* and *Cervus*, is more generally parsimonious and avoids the dubious implication that cervids (deer) are modified bovids.

After the departure of artiodactyls, interpretation of myoglobin sequence data of the remaining eutherians becomes somewhat tenuous. A postulated substitution from histidine to glutamine at locus 81 is the sole synapomorphy in myoglobin (Fig. 3.5: node 58) for *Orycteropus* (aardvark), *Tupaia* (tree-shrew), lagomorphs (rabbits and pikas), a bat (*Rousettus*), hedgehogs, and strepsirhine (lemur-like) primates. But, if this is true, a later back mutation to histidine is required (Figs 3.5 and 3.6: nodes 65, 72, 77 and 82) at locus 81 in *Arctocebus* (Calabar potto), *Callithrix* (marmoset), all Old World anthropoid primates, and the lagomorph *Oryctolagus*. Or, as an equally parsimonious alternative, a back mutation to histidine at locus 81 might characterize both New and Old World anthropoids, followed by a back mutation to glutamine in the New World monkeys *Cebus*, *Saimiri* and *Lagothrix*.

Within this whole assemblage, *Tupaia* and its apparent sister-groups comprising first aardvarks and second lagomorphs, a megachiropteran bat (*Rousettus*) and hedgehogs, are held together weakly (Fig. 3.5: node 59) by a substitution at locus 113 to glutamine, but this substitution occurs also in marsupials, dogs, pigs and Old World anthropoids. Resolution of this part of the cladogram could probably be improved greatly by study of the myoglobin of Dermoptera ('flying lemur'), Rodentia, and additional bats and insectivores. Comparison with the alpha crystallin A chain cladogram is not particularly helpful because of its low resolution with regard to most of these animals, but in the case of the aardvark the myoglobin results are contradicted.

On the basis of the amino acid sequence of myoglobin, primates are united (Fig. 3.6: node 67) by a back mutation to valine at locus 66, requiring a double 'hit'. From that point on, the cladogram results in a set of traditionally accepted relationships based on morphology and is in agreement with the sparser alpha crystallin A chain cladogram. The strepsirhines (lemur-like primates) tested are characterized (Fig. 3.6: node 68) by two synapomorphous substitutions and many autapomorphous ones. Those attributed to *Lepilemur* would doubtless prove to be distributed at various ancestral nodes within the Lemuridae if additional lemurs were to be tested. The haplorhines tested (various monkeys, apes and humans; not yet tarsiers) are widely separated from strepsirhines by five (perhaps six) substitutions in the myoglobin molecule (Fig. 3.6: node 75). Six would be required if the parallel back mutation 81 his−= occurs here rather than in Old World anthropoids only. Among haplorhines, Old and New World anthropoids are characterized by five (perhaps four) substitutions in myoglobin (Fig. 3.6: node 82). The human/chimp/gorilla trio (Fig. 3.6: node 87) that so troubles physical anthropologists, molecular cladists (Darga *et al.*, 1984; Templeton, 1985), and molecular pheneticists (Cronin, Sarich & Ryder, 1984; Sibley & Ahlquist, 1984) is not resolved by sequence data from myoglobin, but synapomorphous relationship of human myoglobin to that of the orang utan is contradicted (cf. Andrews, this volume, p. 36). My results are the same as one of the solutions of Romero-Herrera *et al.* (1976, Fig. 2g) except that the zero-distance problem is handled differently.

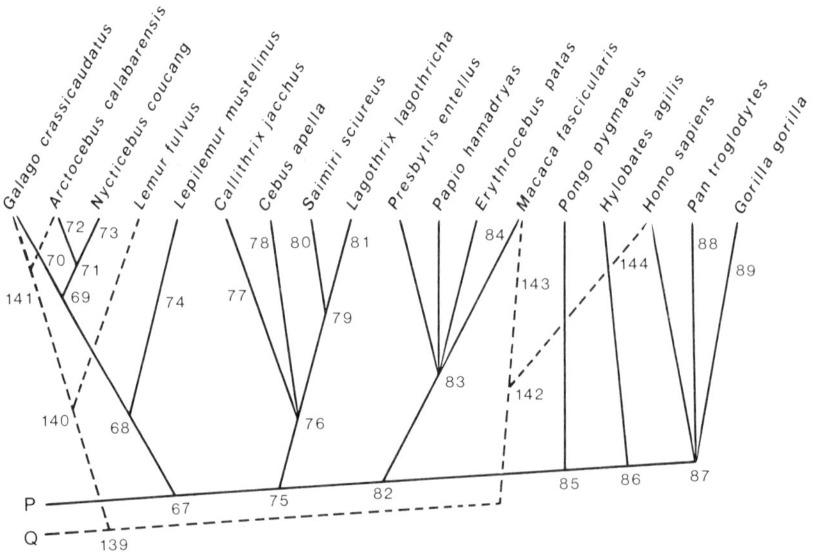

Fig. 3.6 Superposed cladograms of hypothesized phylogenetic changes in tetrapod myoglobin (solid lines) and the alpha crystallin A chain (dashed lines). Part six: primates. P and Q connect to P and Q of Fig. 3.5. See Appendix 1 for explanation of the numbers at the nodes. Common names of genera are: *Galago*, bushbaby; *Arctocebus*, angwantibo; *Nycticebus*, slow loris; *Lemur*, *Lepilemur*, lemurs; *Callithrix*, marmoset; *Cebus*, capuchin monkey; *Saimiri*, squirrel monkey; *Lagothrix*, woolly monkey; *Presbytis*, langur; *Papio*, baboon; *Erythrocebus*, patas monkey; *Macaca*, macaque; *Pongo*, orang utan; *Hylobates*, gibbon; *Homo*, humans; *Pan*, chimpanzee.

Alpha crystallin A chain (lens protein)

A cladogram (Figs 3.1 to 3.6, dashed lines) based on alpha crystallin A chain sequences obtained from living species of 44 genera of mammals plus a frog (*Rana*), a lizard (*Tupinambis*), a crocodilian (*Alligator*), 3 living ratite birds (*Dromaius, Rhea* and *Struthio*), and various other birds was constructed mostly from data supplied through the courtesy of Wilfried de Jong and Morris Goodman. They have already published their interpretation of most of the same data (de Jong & Goodman, 1982; de Jong, 1985). Additional data from extant birds incorporated here are from Stapel *et al.* (1984). I have used the same methods as in constructing the cladogram of myoglobin sequences. Table 3.2 lists the hypothetical sequence of amino acids that I calculate is primitive for the alpha crystallin A chain of Mammalia.

Table 3.2. *Alpha crystallin A chain sequence hypothesized to be primitive for tetrapods (active loci underlined)*

1 met	24 asp	47 tyr	70 lys	93 phe	116 arg	139 leu	162 ser
2 asp	25 gln	48 tyr	71 phe	94 val	117 arg	140 thr	163 arg
3 ile	26 phe	49 arg	72 val	95 glu	118 tyr	141 phe	164 glu
4 thr	27 phe	50 gln	73 ile	96 ile	119 arg	142 ser	165 glu
5 ile	28 gly	51 ser	74 phe	97 his	120 leu	143 gly	166 lys
6 gln	29 glu	52 leu	75 leu	98 gly	121 pro	144 pro	167 pro
7 his	30 gly	53 phe	76 asp	99 lys	122 ser	145 lys	168 thr
8 pro	31 leu	54 arg	77 val	100 his	123 asn	146 val	169 ser
9 trp	32 phe	55 thr	78 lys	101 ser	124 val	147 gln	170 ala
10 phe	33 glu	56 val	79 his	102 glu	125 asp	148 ser	171 pro
11 lys	34 tyr	57 leu	80 phe	103 arg	126 gln	149 asn	172 ser
12 arg	35 asp	58 asp	81 ser	104 gln	127 ser	150 met	173 ser
13 ala	36 leu	59 ser	82 pro	105 asp	128 ala	151 asp	
14 leu	37 leu	60 gly	83 glu	106 asp	129 ile	152 ser	
15 gly	38 pro	61 ile	84 asp	107 his	130 ser	153 ser	
16 pro	39 phe	62 ser	85 leu	108 gly	131 cys	154 his	
17 phe	40 leu	63 glu	86 thr	109 tyr	132 ser	155 ser	
18 tyr	41 ser	64 val	87 val	110 ile	133 leu	156 glu	
19 pro	42 ser	65 arg	88 lys	111 ser	134 ser	157 arg	
20 ser	43 thr	66 ser	89 val	112 arg	135 ala	158 pro	
21 arg	44 ile	67 asp	90 leu	113 glu	136 asp	159 ile	
22 leu	45 ser	68 arg	91 asp	114 phe	137 gly	160 pro	
23 phe	46 pro	69 asp	92 asp	115 his	138 met	161 val	

Although the living taxa whose alpha crystallin A chain has been sampled are not in every case the same as those from which myoglobin has been studied, the cladogram that results is compatible in major topological details with the cladogram of myoglobin evolution. For comparative purposes, I have superposed them in Figs 3.1 to 3.6. As with all cladograms, the characters of the outermost group are in doubt because there is no study of a still more remotely related group. They are either autapomorphies of the outermost group studied, derived from unknown previous character states, or, if symplesiomorphies, they must be modified to yield the characters of the sister-group, or some mix of both extremes must prevail. This unavoidable 'edge effect' dies away as one 'enters' the cladogram.

Basic mammalian divergences

In contrast to the cladogram of myoglobin, only one alpha crystallin A chain synapomorphy (Fig. 3.1: node 92) holds together living reptiles and therian mammals. Presumably, part of the reason for this is that, with the passage of time, once-existing synapomorphies have been converted to various unique autapomorphies, leaving only one for us to know about. Unfortunately, monotremes were not sampled, so the first branch of the Mammalia indicated is that which divides the living Theria into Metatheria and Eutheria (Placentalia), once more as might be expected on morphological grounds. The two extant marsupials sampled are united (Fig. 3.1: node 98) by eight synapomorphous substitutions involving nine 'hits' at the codon level, whereas the Eutheria are held together (Fig. 3.2: node 101) by seven synapomorphous substitutions involving eight 'hits'. As with myoglobin, the 'differentness' of metatherians (marsupials) and eutherians (placentals) from their common ancestor is about the same in each case, yet in the world of 'morphological differentness' it is still common practice to regard the metatherians as representing one order, but to assign the eutherians to 30 or so, as noted above. The Australian kangaroo *Macropus rufus* has two autapomorphous substitutions and the American opossum *Didelphis marsupialis* five, in keeping with long isolation.

In contrast to the list of animals represented in the cladogram of myoglobin sequences, four edentates (two anteaters and the two living, distantly related sloths) appear in the cladogram of the alpha crystallin A chain. They, together with elephant shrews (*Elephantulus*), paenungulates and the rest of the living eutherians studied, form a fourfold array (Fig. 3.2: node 101) that cannot be resolved on the basis of the alpha crystallin A chain. In the various cladograms presented by Goodman *et al.* (1985 and elsewhere) edentates are presented as the sister-group of Proboscidea, an interesting concept for which few anatomists are yet ready! Of the branches of the fourfold split hypothesized in this paper, that represented by the extant edentates (Fig. 3.2: node 102) has nine synapomorphous substitutions or deletions of its own before branching into the divisions represented by sloths (node 103) and anteaters (node 106). In addition to the line leading to the edentates, one leads without known autapomorphous substitutions to the partially sequenced elephant shrew *Elephantulus* (de Jong, 1985), a third one goes to various living aardvarks and paenungulates (African elephant, hyrax and manatee) (Fig. 3.2: node 108), and a fourth leads to the remaining eutherians (Fig. 3.3: node 113). De Jong, Sweers & Goodman (1981) concluded that, on the basis of the sequence of amino acids in its alpha crystallin A chain, the aardvark belongs in the paenungulate group. On serological grounds, J. Shoshani, M. Goodman & W. Prychodko (unpublished data) infer that elephants, sirenians, hyracoids and aardvarks are related to one another. Shoshani *et al.* (1985) also infer that, on the basis of studies of haemoglobin, the hyrax is a paenungulate. The Proboscidea thus fall in the same genealogical position in the lens protein pattern as they do in the cladogram of myoglobin. Their immediate sister-group among living mammals is a threefold array (Fig. 3.2: node 110), comprising *Orycteropus* (aardvark), *Procavia* (hyrax) and *Trichechus* (manatee). However, analysis of the myoglobin sequence of the aardvark contradicts this position for it.

No matter what the details, many molecular cladograms place the origin of paenungulates at a very ancient point of subdivision within the Eutheria, about as ancient as the origin of the living elephant shrews and possibly the Edentata, although not necessarily indicating that such animals as anteaters and elephants are more closely related to one another than to other extant eutherians.

Higher eutherian divergences

After the branching away of the lineages leading to Macroscelidea, aardvarks, paenungulates and edentates, all the other living eutherian mammals in which the alpha crystallin A chain has been sequenced fall into a monophyletic but unfortunately poorly resolved group (Fig. 3.3: node 113). This group is characterized by three synapomorphous substitutions: at locus 91, aspartic acid is replaced by glutamic acid, locus 150 changes from methionine to leucine, and locus 153 shifts from serine to glycine. Several alternative schemes can be devised to explain the distribution of amino acids among the alpha crystallin A chains of the members of this fourfold group, whose constituents are *Erinaceus* (hedgehog), *Tupaia* (treeshrew), an alliance of rodents, lagomorphs and primates, and another alliance consisting of carnivores, pangolins and the ungulate orders (whales, perissodactyls, and artiodactyls). The microchiropteran Caribbean fruit bat *Artibeus* splits away from the base of this last alliance (Fig. 3.3: node 114). An alternative scheme is given by de Jong (1985, Fig. 2), and there is an equally parsimonious variant that substitutes 90 Q–L (glutamine to leucine, see Appendix 2) in myomorph rodents (cf. Fig. 3.5: node 138) in place of 90 L–Q in two hystricognath rodents.

Carnivores and ungulates (Simpson's Ferungulata, 1945) and pangolins are characterized by two synapomorphous substitutions (Fig. 3.3: node 116) in living representatives. Addition of soricoid insectivores (shrews) to the sample tested would possibly help to resolve the position of the Caribbean fruit bat (*Artibeus*), which at first glance seems out of place in the present scheme. However, the position of *Artibeus* may imply merely that microchiropteran bats had their origin in the broad palaeoryctoid 'insectivore' radiation of late Cretaceous times. Unfortunately, analytical techniques at present would require a very large number of shrews in order to compile enough of the alpha crystallin A chain to compare with *Artibeus* and other mammals. Studies of other molecules place the bats as the sister-group of 'insectivores' (Goodman *et al.*, 1985).

After the phylogenetic departure of the Caribbean fruit bat and the acquisition of two synapomorphous substitutions at loci 4 and 147 (Fig. 3.3: node 116), an unresolved sixfold fragmentation occurs. Four of the fragments lead to living carnivorans, but little useful information about carnivoran interrelationships is provided by the alpha crystallin A chain. A fifth line leads, after two further synapomorphous substitutions (Fig. 3.3: node 120), to the two pangolins (scaly anteaters, Pholidota) tested, *Manis* and *Phataginus*. (Actually, one NR can be saved by making the two pangolins the sister-group of the sloth bear, *Melursus*. They share a mutation to tyrosine at locus 74.) This molecular result roughly fits the conclusions reached recently by serologists (J. Shoshani, M. Goodman & W. Prychodko, unpublished data) and by some palaeontologists to the effect that living pangolins and their fossil palaeanodont allies of the early Caenozoic of the Northern Hemisphere originated from early pantolestid-like 'insectivores' (Rose, 1978) that were themselves closely related to early carnivorans and Cretaceous palaeoryctoids, but special relationship to *Melursus* verges on the incredible. A relationship of living pangolins to extant edentates must be more remote, possibly involving a latest common ancestor that lived well back in Cretaceous time, prior to the acquisition by pholidotan ancestors of the seven substitutions that characterize nodes 113, 114 and 116.

The sixth line (Fig. 3.4: node 123) leads to the cetaceans plus the traditional ungulates (Perissodactyla and Artiodactyla). A single, parallel-prone substitution at locus 90 holds the group together, but, this time, in contrast to the cladogram of myoglobin, the Perissodactyla join the Artiodactyla (Fig. 3.5: node 126) rather than the whales. At least this situation is not as embarrassing as that of cytochrome *c*, wherein perissodactyls and prawns group together on the basis of low NR scores (Morris Goodman, personal communication)! Within the Perissodactyla, traditional relationships are supported (nodes 127, 128), but in the Artiodactyla, the alpha crystallin A chain is of no help yet in resolving relationships because no synapomorphous substitutions occur. Moreover, in strong contrast to the cladogram based upon myoglobin, it is *Sus* this time that has acquired three times as many autapomorphies as its nearest competitor among the euartiodactyls tested.

The remaining mammals (from node 113) represent the orders Lagomorpha (rabbits and pikas), Rodentia and Primates. Their relationships cannot be resolved well yet (Fig. 3.5: node 135). As noted above, other, equally parsimonious schemes are possible, given the small amount of available data. However, such schemes either involve unlikely conclusions, such as that the laboratory rabbit is closer to primates than it is to *Ochotona* (pika), or indicate that the lagomorphs are more closely related to the hedgehog *Erinaceus* than to members of the order Primates.

None of these alternative molecular schemes either supports or denies the idea that the Lagomorpha and Rodentia (currently classified together in cohort Glires) are more closely related to each other than to other eutherian groups (see also Penny & Hendy, 1985*a*, p. 81). However, morphological evidence from fossils of the family Eurymylidae (Rodentia from a cladistic viewpoint) of the early Caenozoic of Asia have been interpreted by some workers (e.g. Li & Ting, 1985; Novacek, 1985) to indicate that rabbits and rodents share a common ancestor in the Palaeocene or Cretaceous of Asia. Only an increase in richness of diversity sampled, both of proteins and of fossils, will improve this unsatisfactory situation significantly. In view of the apparent slowness of alpha crystallin A chain evolution in rodents, other proteins should be turned to. Already, Beintema (1985) unites African and South American hystricognath rodents on the basis of studies of pancreatic ribonuclease and possibly also insulin. Myoglobin of several rodents has recently been sequenced and suggests that rodents and rabbits are closely related (D. Penny, personal communication).

Primates as a whole are held together by a single substitution (Fig. 3.6: node 139) in the alpha crystallin A chain, but six synapomorphous substitutions (Fig. 3.6: node 142) unite the two anthropoids tested. In alternate schemes, this number is either five or seven, but it is impressive in any case.

General remarks on molecular cladograms based on sequence data

In the process of constructing the cladograms of myoglobin and alpha crystallin A chain amino acid sequences, one is struck by the amount of parallel and back mutation that must be invoked in order to keep schemes as parsimonious as possible. This is not to say that nature is always parsimonious, but to assume that it is generally so is reasonable because it permits analysis, whereas to assume non-parsimony *a priori* leads to an abdication of rationality.

How do amino acid substitutions become pervasive throughout a population? Surely, change is not instantaneous throughout a whole population; rather, frequencies of occurrence of various possible amino acids at any particular active locus in a polymorphic molecule shift in time from low to high or high to low, either slowly or rapidly depending on whether the frame of reference is the time scale of ordinary human experience or that of the palaeontological record. For this reason, it seems to me probable that some of the polypeptide sequences determined thus far from single specimens of a species are not necessarily typical of whole populations. Therefore, 'corrections' made in new investigations of the same polypeptide from additional individuals of a species may in some cases prove to be polymorphism rather than improvements upon incorrect previous sequencing. Time will tell.

Although the sets of living animals in which myoglobin and the alpha crystallin A chain have been studied are somewhat different, considerable overlap exists (Figs 3.1 to 3.6). In major taxa, both molecules give essentially the same cladistic topology or they are merely unhelpful because of lack of synapomorphous substitutions capable of resolving some particular phylogenetic issue. Similar results for five different molecules were obtained by Penny, Foulds & Hendy (1982) and by Penny & Hendy (1985*b*).

Both molecular cladograms presented here cry out for amplification. Presumably, this shortcoming will be rectified as taxonomically more appropriate animals, in addition to those merely easily available, are tested. Such additions will not only increase the diversity of the cladograms constructed, but also force them to be modified in many ways and will undoubtedly reverse some of the polarities that are currently accepted in attempts to achieve parsimony. In groups such as rodents, bats, carnivorans, 'insectivores' and artiodactyls, there is clearly much work to do before the most parsimonious branching structures become reasonably clear. Even the place of the Mammalia among the vertebrates can be seen more clearly by adding additional outgroups (cf. Bishop & Friday, this volume, p. 128; Goodman, Miyamoto & Czelusniak, this volume, p. 151). The alpha crystallin A chain needs to be studied in monotremes and in a variety of non-mammalian tetrapods in addition to those studied thus far: one frog (*Rana*), a crocodilian (*Alligator*), a lizard (*Tupinambis*, the tegu), and a scattering of birds (Stapel *et al.*, 1984).

Additional molecules, such as phospholipases and lipotropins, need to be studied from a phylogenetic standpoint.

Palaeontology and comparative anatomy

Phylogenetic analysis of the relationships of most eutherian mammalian taxa at the ordinal level is still the subject of much seesawing debate, more than 200 years after Linnaeus. Thus, if one looks long enough in the literature, one can find many suggestions. Some of these are highly unlikely, such as a proposed linkage of lagomorphs with bats or an alliance of rodents with edentates (Shoshani *et al.*, 1985).

Cladistic analysis of the morphology of mammals is still very much in its infancy, but gradually the situation is improving as more animals and more characters and character complexes are incorporated into ever more detailed cladograms. Preferably, some sort of functional context should be provided for characters and character complexes as Szalay (1977, 1985) insists, but analysis of distribution of many characters among many organisms can provide excellent results even without a thorough knowledge of function in either living organisms or fossils. Knowledge of function is always a matter of interpretation in fossils.

Cladistic analysis of mammal-like tetrapods

In order to put the Mammalia in perspective, I have prepared a cladogram (Fig. 3.7; Appendix 3) of some basic features of mammalian and premammalian morphology taken unashamedly from stimulating papers by Kemp (1983) and Sues (1985). I have modified their conclusions slightly, notably with regard to the position occupied by the multituberculates. Multituberculates are a highly successful and diverse group of extinct herbivorous mammals superficially similar to rodents. Multituberculates lived on northern continents from Norian times, at the end of the Triassic, until the beginning of Oligocene time. The cladogram reflects most of what Kemp had to say, and it makes predictions about what sort of anatomy can be expected as Mesozoic mammals become better known. For instance, with regard to hard parts likely to be fossilized, it predicts that monotremes will prove to have evolved from ancestors with teeth forming reversed triangles. The position of the living monotremes is that of the sister-group of the living Theria, but a number of extinct Mesozoic stem-groups fall between the mammalian crown-group (monotremes, marsupials and eutherians) and the extinct stem-group generally thought of as 'therapsids' (see Jefferies, 1979, and Ax, 1984, for discussion of crown- and stem-groups). From the cladogram, one can see that the definition of Mammalia is arbitrary: one could place the limit of the Mammalia at any of the synapomorphous nodes except the one represented by character 68, leading to such opposite extremes as the mammalian definitions of Reed (1960) and Van Valen (1960) or MacIntyre (1967). Neontologists (a palaeontological epithet) would no doubt be happy to restrict the Mammalia to the living and fossil therians and monotremes (characters 56 and 60 to 67), and MacIntyre (1967), in company with early naturalists like Ray and Linnaeus, would have them confined to just the marsupials and placentals (characters 70 to 75). Palaeontologists will surely continue to squabble over whether this or that relatively plesiomorphous monophyletic fossil group should be included in the Mammalia on the basis of various arbitrary definitions (Simpson, 1959, 1960). One thing is clear, however: terms like 'cynodonts', 'eucynodonts', 'advanced cynodonts', 'therapsids', 'quasi-mammals' and 'mammal-like reptiles' are expressions of paraphyly and are the result of subtracting one monophyletic group from another. Elsewhere (McKenna, 1982) I have suggested that many of these could more logically be referred to as 'reptile-like mammals'. Gosliner & Ghiselin (1984) and Ax (1985) agree. However, I do not find any of these terms very useful.

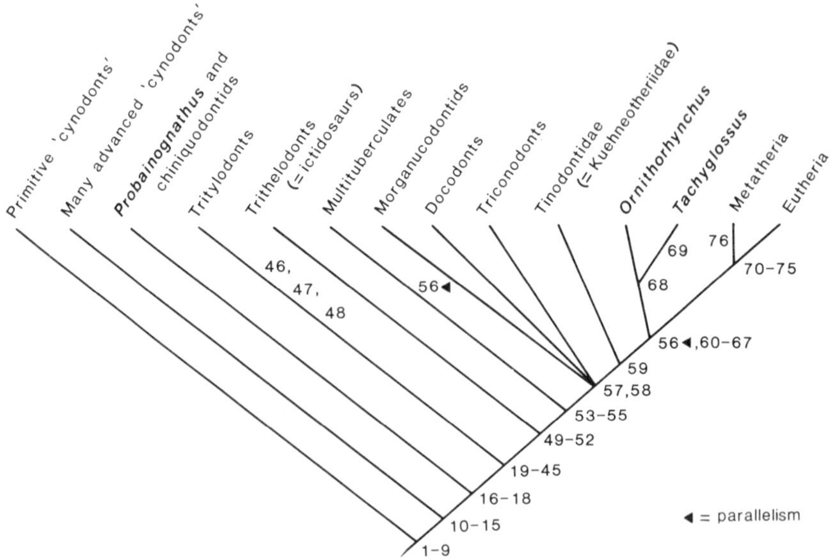

Fig. 3.7 Cladogram of mammals and some reptile-like stem-groups. See Appendix 3 for explanation of the numbers at the nodes.

From Kemp's (1983) work and its cladistic analysis here, it can be concluded that the living monotremes are not some long-hidden, very early Mesozoic group that resurfaced, *Latimeria*-like, in Australia at a late date. Instead, they could have split away from primitive Theria at any time up to about the beginning of the Cretaceous, possibly not even in Australia, although it is reasonable to think of them as an Australian product. The fossil history of monotremes has indeed until recently been a hidden one, but the Australian middle Miocene (22 MyrBP) genus *Obdurodon* (Woodburne & Tedford, 1975) partially filled the long gap. *Steropodon*, a recently-found early Cretaceous (Albian) monotreme from Australia (Archer *et al.*, 1985), extends the known range of fossil monotremes back to 107 MyrBP. The lower dentition of *Steropodon* closely resembles that of tribosphenic therians. Therefore, monotremes apparently did not branch away from the stem leading to metatherians and eutherians very long before they themselves diverged as separate lineages. Thus, if change is partially correlated with the passage of time, the recently revealed palaeontological record of monotremes is in accord with the molecular information, because relatively few NRs occurred between the departure of the monotremes from the therian stem and the fragmentation of that stem into marsupials and placentals. It is not in accord with earlier hypotheses that unite monotremes with ancient multituberculates partly on the basis of symplesiomorphous characters (Kielan-Jaworowska, 1971).

Edentata (Xenarthra)

Simpson (1945, p. 190) was of the opinion that Edentata originated at about the beginning of the Palaeocene, some 65 MyrBP, from what he called proto-Insectivora. Simpson's edentates included not only the xenarthrous sloths, anteaters and armadillos, mainly South American mammals some of whose vertebrae exhibit extra articulations with each other (the xenarthrous condition), but also a non-xenarthrous (nomarthrous) and extinct group known as the Palaeanodonta. Palaeanodonts are known from northern continents (only North America when Simpson studied them). Xenarthrans clearly were isolated for a long time in South America before spreading with varying success northward, after North and South America neared one another in the late Caenozoic. How long xenarthrous edentates were cooped up in South America before their late Caenozoic release is unknown, but Simpson believed that before their incarceration in South America they had an ultimately North American origin from ancestors that resembled known palaeanodonts of the Eocene. Szalay & Decker (1974) and Szalay (1977) support the same view.

There is a problem with this scenario: the palaeanodonts have been claimed to be more closely allied to the pangolins than to xenarthrous edentates (Emry,

1970; see Szalay, 1977, and Rose, 1979, for disagreement). Moreover, in 1975, I pointed out that xenarthrous edentates seemed to have a poorly differentiated uterus and vagina, lacking a definite cervix (Forbes, 1882), which, along with several other characters, indicated that xenarthrous edentates must be a very early eutherian branch that departed before the cervix came to characterize the rest of the Eutheria. I saw no reason why the xenarthrous edentates could not have been isolated in South America since the final break-up of Pangaea, when dry land between Africa and South America was finally sundered in the late Cretaceous. They could have been in existence even before that. The molecular cladograms worked out by many workers, plus that for the alpha crystallin A chain presented here, are in accord with this view.

On the other hand, the palaeanodonts and pangolins do seem to tie together with a family of 'proto-Insectivora', an extinct paraphyletic group that Romer (1966) and others have called 'Proteutheria', namely the family Pantolestidae (Szalay, 1977; Rose, 1978), whose roots in the Late Cretaceous lie close to those of the Carnivora and thus are in general accord with the molecular cladogram of the alpha crystallin A chain given here. Recently, palaeanodonts described from both China and Europe (Ding, 1979; Storch, 1981) have been identified as edentates, but, in spite of their authors' claims, I do not believe that they have been demonstrated to be xenarthrous (see also Novacek, 1982). New evidence bearing on the origin of the pangolins is currently under study by R. J. Emry.

The Glires problem

Recently, Novacek (1985) has compiled a table of 36 morphological characters distributed among the Rodentia and 12 major candidates for the sister-group of the Rodentia. Various other mammalian groups were omitted. His results are depicted in a simplified cladogram (1985, fig. 5) that shows the Lagomorpha to be the most likely candidate (among those considered) as sister-group of the Rodentia. Rodents and lagomorphs are together placed in a higher taxon, the cohort Glires.

However, if one takes into account foot structure (Szalay, 1985), any close relationship of rodents with lagomorphs must imply either rapid evolution of the early lagomorph astragalus and calcaneum away from a more primitive foot structure slightly modified by rodents, or a long period of evolution after the phylogenetic separation of the ancestors of lagomorphs from the ancestors of rodents (A. Bleefeld, in preparation). If a long period of evolution is correct, this in turn would push back into the Cretaceous the branch point between Glires and their sister-group, which may be 'Archonta', a group including primates and bats.

That Glires is natural is not questioned, but it is not natural by reason of the enlarged incisors that give the group its name. This is because of the incisors of the family Pseudictopidae of the Asian early Caenozoic (Sulimski, 1969). Pseudictopids are lagomorphs in both dental pattern and in the structure of the foot; however, there is no enlargement of any particular incisor. Therefore, early lagomorphs probably did not have enlarged incisors (a reversal to subequal incisors in pseudictopids seems unlikely). This, in turn, means that the enlarged gliriform incisors of rodents and lagomorphs are convergent. It is of further interest that the incisors of *Pseudictops* are multicuspate, somewhat resembling those of primitive dermopterans (Rose, 1982).

Macroscelidea

The Macroscelidea (elephant shrews) have always been considered to be endemic African mammals, possibly distantly related to tupaiids, primates or insectivores, but not demonstrably related to any of these on the basis of shared–derived morphological characters (Evans, 1942; Corbet & Hanks, 1968; Novacek, 1984). They have been given their own order by Butler (1956), but that hardly solves the question of their genealogical relationships. They may be related to the extinct Asian family Anagalidae (Evans, 1942; McKenna, 1975). The known fossil record of macroscelideans goes back to the African Oligocene and perhaps as far back as the African early Eocene (Hartenberger, 1986) (or to the Chinese Palaeocene if anagalids are truly involved). Our ignorance of macroscelidean relationships might be remedied by a thorough restudy of the skeleton of the early Oligocene genus *Anagale* from Mongolia. In the meantime, it is difficult to dispute the molecular evidence from the alpha crystallin A chain that the Macroscelidea have long been separate from other mammals (de Jong, 1985). Their fossil history in Africa has been discussed by Butler & Hopwood (1957), Patterson (1965), Butler (1969, 1978), Butler & Greenwood (1976), and Novacek (1984).

Paenungulata

Simpson (1945) first defined the Paenungulata as a superorder of his cohort Ferungulata. The other ferungulate superorders were Ferae (carnivorans and the extinct creodonts, the latter including arctocyonids and mesonychids), Protungulata (tubulidentates and the extinct litopterns, notoungulates, astrapotheres, and condylarths minus the arctocyonids), Mesaxonia (perissodactyls), and Paraxonia (artiodactyls). The Paenungulata originally comprised the seven orders Pantodonta, Dinocerata, Pyrotheria, Embrithopoda, Proboscidea, Hyracoidea and Sirenia. The first four of these seven orders are known only from fossils. Sirenia included the extinct Desmostylia. Various degrees of doubt apply to the inclusion of these fossil groups in a natural (monophyletic) Paenungulata, but the Sirenia, Proboscidea and the by-then separate order Desmostylia were united in a group called Tethytheria by McKenna (1975). The naturalness of this smaller group, plus a looser association of tubulidentates, hyracoids and probably the extinct embrithopods, has been championed by various workers since then, notably de Jong *et al.* (1981), J. Shoshani, M. Goodman & W. Prychodko (unpublished data), and Domning, Ray & McKenna (1986).

Simpson (1945) placed the aardvarks (*Orycteropus*, order Tubulidentata) as the only living members of a plesiomorphous and paraphyletic 'group' that he called Protungulata. The meaning of this was little more than that he regarded aardvarks as primitive. He presumed them to be allied to primitive 'ungulates' and therefore to 'condylarths'. The latter are themselves a scrap basket of extinct paraphyletic groups. Patterson (1975), the authority on fossil aardvarks, did not present an analysis of shared–derived characters that would link aardvarks with any particular monophyletic mammalian group, although he did present a history of other people's ideas on the matter and noted Sonntag's (Sonntag, 1925; Clark & Sonntag, 1926) work favourably comparing the anatomy of *Orycteropus* with that of elephants and hyraxes. 'Condylarths' were at that time considered to be ancestral to all 'ungulates', including elephants and hyraxes. The latest study of tubulidentate anatomy (Thewissen, 1985) also suggests that aardvarks are more primitive than (true) ungulates. However, Thewissen confuses paenungulates (subungulates) with true ungulates when he states (1985, p. 280) that his conclusions are in conflict with an assignment of the aardvarks to the Paenungulata. Thus, there is at present no palaeontological or anatomical objection to placing aardvarks in the phylogenetic position suggested by the cladogram of the alpha crystallin A chain presented here (Fig. 3.2) or that by de Jong *et al.* (1981). However, these results contradict those derived from the study of myoglobin.

Although there are some similarities to perissodactyls in the dentition of the early Caenozoic embrithopod-like family Phenacolophidae and some similarities as well in the Hyracoidea (for instance as claimed by Van Valen, 1971, p. 424), there are no convincing morphological links between the tethytheres and the true 'ungulates' (artiodactyls, perissodactyls and whales). The earliest known tethytheres occur in the Palaeocene of China: *Minchenella*, a suitable ancestor for both the living Proboscidea and the extinct Desmostylia. The living order Sirenia, whose first occurrences in the record are in the Eocene, appears to be the sister-group of a group comprising the proboscideans, desmostylians and *Minchenella* (Domning *et al.*, 1986). Probably, the living hyracoids and fossil embrithopods (arsinoitheres) fit in as well.

Primitive Sirenia still possessed five premolars, both deciduous and permanent (replacement) ones (Domning, Morgan & Ray, 1982; Domning *et al.*, 1986). Why is this important?

More than 10 years ago (McKenna, 1975), I published a paper in which I pointed out that all was not well regarding the time-honoured eutherian dental formula. This, as in marsupials, called for seven postcanine teeth as the primitive condition. But Mesozoic eutherians were turning up with eight: five premolar loci and three molar loci. If a count of five premolars is primitive, how did the number get reduced to four in various Caenozoic eutherian groups? Was the method always the same?

The Mesozoic eutherians with eight postcanine teeth include the early Cretaceous *Prokennalestes*, which resembles the leptictids, early Caenozoic insectivores. In the late Cretaceous, there are leptictid relatives (*Kennalestes*, *Gypsonictops*) that seem to have been losing the tooth at the third premolar locus. In the Caenozoic leptictids, the postcanine count is seven, but the remaining premolar loci are P1, P2, P4, P5; and the molars are M2, M3, M4 in the numbering scheme used for living mammals. Deciduous and permanent P5s resemble each other, but that is not the case at the P4 locus, a clue that aids in identification.

No one knows how the reduction from eight to seven postcanine teeth took place in non-leptictid eutherians. My own explanation for that loss amounted to a postulation of paedomorphy: retention of a deciduous tooth at the fifth premolar locus (dP5) to produce a new, unreplaced first molar, and non-development of the last molar in more advanced forms to return to a condition with only three molars (dP5 and two molars behind it, known traditionally as M1, M2, M3). Except for its explanation of leptictids, my hypothesis has not been generally accepted, nor has it been clearly refuted.

That the ancestors of all known Caenozoic eutherian mammals (except possibly the edentates) had reduced their postcanine dental formula by some means to seven teeth by the end of the Cretaceous was considered secure until recently, when studies of Eocene sirenians by Domning *et al.* (1982) showed conclusively that the primitive condition in early Sirenia, like that of primitive Cretaceous eutherians, was still to possess eight postcanine loci, of which five were for replaced premolars. Thus, unless this retention represents some sort of most exceptional reversal in evolution, the Sirenia (and possibly the ancestors of their tethytherian and other paenungulate allies) represent an early side-branch (or branches) of the Eutheria that, at the time of branching away, had not yet attempted a reduction in the number of their postcanine teeth. This, in turn, implies an early event at which the paenungulates became separated from other eutherians, retaining their primitive postcanine dental formula in Sirenia until the Eocene. This event must have occurred sometime in the Cretaceous, well before the origin of true ungulates like artiodactyls or perissodactyls.

This outcome is in agreement, once more, with results from studies of myoglobin and the alpha crystallin A chain.

Conclusions

As in murder investigations, reconstruction of phylogenetic history after the event has its difficulties. There may be several versions of what supposedly has happened. Nevertheless, only one sequence of events actually occurred. As with murder trials, we may not be absolutely sure of a particular version of history, but we can be sure beyond reasonable doubt in many cases, especially when a number of disparate lines of evidence point to the same general result. Most techniques of phylogenetic reconstruction are beset by difficulties, but comparison of results from various disciplines as remote as palaeontology and protein sequencing are seen here to yield essentially the same phylogenetic reconstructions. As with all information, there is a mixture of signal and noise, such as that associated with aardvarks and possibly also pangolins, but the situation seems to be becoming quieter.

ACKNOWLEDGEMENTS

It is a pleasure to thank Wilfried de Jong, Morris Goodman, Howard Dene, Alejo Romero-Herrera, Jeheskel Shoshani, and especially Colin Patterson for their help, patience and generosity. Michael Novacek, Larry Flynn, Ann Bleefeld and André Wyss provided thoughtful criticism and information. The authorities in charge of the University of Colorado's high altitude research station permitted me to collect enough pikas for de Jong to determine the amino acid sequence of *Ochotona* alpha crystallin A chain and for Dene and Romero-Herrera to discover the amino acid sequence of its myoglobin.

REFERENCES

Archer, M., Flannery, T. F., Ritchie, A. & Molnar, R. E. (1985). First Mesozoic mammal from Australia – an early Cretaceous monotreme. *Nature, London*, **318,** 363–6.

Ax, P. (1984). *Das phylogenetische System.* Stuttgart: Fischer.

Ax, P. (1985). Stem species and the stem lineage concept. *Cladistics*, **1**(3), 279–87.

Beintema, J. J. (1985). Amino acid sequence data and evolutionary relationships among hystricognaths and other rodents. In *Evolutionary Relationships among Rodents: a Multidisciplinary Analysis*, ed. W. P. Luckett & J.-L. Hartenberger, pp. 549–65. New York: Plenum Publishing Corporation.

Benton, M. J. (1985). Classification and phylogeny of the diapsid reptiles. *Zoological Journal of the Linnean Society*, **84,** 97–164.

Butler, P. M. (1956). The skull of *Ictops* and the classification of the Insectivora. *Proceedings of the Zoological Society of London*, **126,** 453–81.

Butler, P. M. (1969). Insectivores and bats from the Miocene of East Africa: new material. In *Fossil Vertebrates of Africa*, vol. 1, ed. L. S. B. Leakey, pp. 1–38. New York & London: Academic Press.

Butler, P. M. (1978). Insectivora and Chiroptera. In *Evolution of African Mammals*, ed. V. J. Maglio & H. B. S. Cooke, pp. 56–68. Cambridge, Massachusetts, & London: Harvard University Press.

Butler, P. M. & Greenwood, M. (1976). Elephant-shrews (Macroscelididae) from Olduvai and Makapansgat. In *Fossil Vertebrates of Africa*, ed. R. J. G. Savage & S. C. Coryndon, vol. 4, pp. 1–56. London: Academic Press.

Butler, P. M. & Hopwood, A. T. (1957). Insectivora and Chiroptera from the Miocene rocks of Kenya Colony. *Fossil Mammals of Africa*, no. 13. London: British Museum (Natural History).

Carpenter, J. M. (1985). The clock of evolution needs repair. *Natural History*, **94**(6), 6.

Clark, W. E. Le Gros & Sonntag, C. F. (1926). A monograph of *Orycteropus afer*. III. The skull; the skeleton of the trunk and limbs; general summary. *Proceedings of the Zoological Society of London*, **1926**, 445–85.

Corbet, G. B. & Hanks, J. (1968). A revision of the elephant-shrews, family Macroscelididae. *Bulletin of the British Museum (Natural History), Zoology*, **16,** 47–111.

Crick, F. H. C. (1958). On protein synthesis. *Symposia of the Society for Experimental Biology*, **12,** 138–63.

Cronin, J. E., Sarich, V. M. & Ryder, O. (1984). Molecular evolution and speciation in the lesser apes. In *The Lesser Apes, Evolutionary and Behavioural Biology*, ed. H. Preuschoft, D. J. Chivers, W. Y. Brockelman & N. Creel, pp. 467–85. Edinburgh University Press.

Darga, L. L., Baba, M. L., Weiss, M. L. & Goodman, M. (1984). Molecular perspectives on the evolution of the lesser apes. In *The Lesser Apes, Evolutionary and Behavioural Biology*, ed. H. Preuschoft, D. J. Chivers, W. Y. Brockelman & N. Creel , pp. 448–66. Edinburgh University Press.

de Jong, W. W. (1985). Subordinal affinities of Rodentia studied by sequence analysis of eye lens protein. In *Evolutionary Relationships among Rodents: a Multidisciplinary Analysis*, ed. W. P. Luckett & J.-L. Hartenberger, pp. 211–26. New York: Plenum Publishing Corporation.

de Jong, W. W. & Goodman, M. (1982). Mammalian phylogeny studied by sequence analysis of the eye lens protein alpha-crystallin. *Zeitschrift für Säugetierkunde*, **47**(5), 257–76.

de Jong, W. W. & Rydén, L. (1981). Causes of more frequent deletions than insertions in mutations and protein evolution. *Nature, London*, **290,** 157–9.

de Jong, W. W., Sweers, A. & Goodman, M. (1981). Relationship of aardvark to elephants, hyraxes and sea cows from alpha-crystallin sequences. *Nature, London*, **292,** 538–40.

Ding, S.-Y. (1979). A new edentate from the Paleocene of Guangdong. *Vertebrata Palasiatica*, **17**(1), 57–64.

Domning, D. P., Morgan, G. S. & Ray, C. E. (1982). North American Eocene sea cows (Mammalia: Sirenia). *Smithsonian Contributions to Paleobiology*, **52,** 1–69.

Domning, D. P., Ray, C. E. & McKenna, M. C. (1986). Two new Oligocene desmostylians and a discussion of tethytherian systematics. *Smithsonian Contributions to Paleobiology*, **59,** i–iii, 1–56.

Emry, R. J. (1970). A North American Oligocene pangolin and other additions to the Pholidota. *Bulletin of the American Museum of Natural History*, **142,** 455–510.

Evans, F. G. (1942). The osteology and relationships of the elephant-shrews (Macroscelididae). *Bulletin of the American Museum of Natural History*, **80,** 85–125.

Felsenstein, J. (1978). The number of evolutionary trees. *Systematic Zoology*, **27,** 27–33.

Fermi, G. & Perutz, M. F. (1981). *Atlas of Molecular Structures in Biology. 2. Haemoglobin and Myoglobin.* Oxford: Clarendon Press.

Forbes, W. A. (1882). On some points in the anatomy of the great anteater (*Myrmecophaga jubata*). *Proceedings of the Zoological Society of London*, **1882,** 287–302.

Gardiner, B. G. (1982). Tetrapod classification. *Zoological Journal of the Linnean Society*, **74**(3), 207–32.

Goodman, M., Czelusniak, J. & Beeber, J. E. (1985). Phylogeny of primates and other eutherian orders: a cladistic analysis using amino acid and nucleotide sequence data. *Cladistics*, **1**(2), 171–85.

Goodrich, E. S. (1916). On the classification of the Reptilia. *Proceedings of the Royal Society, B*, **89,** 261–76.

Gosliner, T. M. & Ghiselin, M. T. (1984). Parallel evolution in opisthobranch gastropods and its implications for phylogenetic methodology. *Systematic Zoology*, **33,** 255–74.

Graham, R. L. & Foulds, L. R. (1982). Unlikelihood that minimal phylogenies for a realistic biological study can be constructed in reasonable computational time. *Mathematical Biosciences*, **60,** 133–42.

Gregory, W. K. (1947). The monotremes and the palimpsest theory. *Bulletin of the American Museum of Natural History*, **88,** 1–52.

Hartenberger, J.-L. (1986). Hypothèse paléontologique sur l'origine des Macroscelidea (Mammalia). *Compte Rendu Hebdomadaire des Séances de l'Académie des Sciences, Paris*, **32,** series II, 5, 247–9.

Hull, D. L. (1979). The limits of cladism. *Systematic Zoology*, **28**(4), 416–40.

Jefferies, R. P. S. (1979). The origin of chordates – a methodological essay. In *The Origin of Major Invertebrate Groups*, ed. M. R. House, pp. 443–77. London & New York: Academic Press.

Kemp, T. S. (1983). The relationships of mammals. *Zoological Journal of the Linnean Society*, **77,** 353–84.

Kielan-Jaworowska, Z. (1971). Skull structure and affinities of the Multituberculata. *Palaeontologica Polonica*, **25,** 5–41.

Li, C.-k. & Ting, S.-y. (1985). Possible phylogenetic relationship of Asiatic eurymylids and rodents, with comments on mimotonids. In *Evolutionary Relationships among Rodents: a Multidisciplinary Analysis*, ed. W. P. Luckett & J.-L. Hartenberger, pp. 35–58. New York: Plenum Publishing Corporation.

MacIntyre, G. T. (1967). Foramen pseudovale and quasi-mammals. *Evolution*, **21**(4), 834–41.

McKenna, M. C. (1975). Toward a phylogenetic classification of the Mammalia. In

Phylogeny of the Primates: an Interdisciplinary Approach, ed. W. P. Luckett & F. S. Szalay, pp. 21–46. New York: Plenum Press.

McKenna, M. C. (1982). More of dinosaur extinction, and more. (Review of *Hunting the Past: Fossils, Rocks, Tracks and Trails – the Search for the Origin of Life*, by L. B. Halstead.) *Nature, London*, **300,** 560.

Moore, G. W. (1975). Proof of the maximum parsimony ('Red King') algorithm. In *Molecular Anthropology, Genes and Proteins in the Evolutionary Ascent of the Primates*, ed. M. Goodman, R. E. Tashian & J. H. Tashian, pp. 117–37. New York & London: Plenum Press.

Nelson, G. & Platnick, N. I. (1981). *Systematics and Biogeography: Cladistics and Vicariance*. New York: Columbia University Press.

Novacek, M. (1982). Information for molecular studies from anatomical and fossil evidence on higher eutherian phylogeny. In *Macromolecular Sequences in Systematic and Evolutionary Biology*, ed. M. Goodman, pp. 3–41. New York: Plenum Publishing Corporation.

Novacek, M. (1984). Evolutionary stasis in the elephant-shrew, *Rhynchocyon*. In *Living Fossils*, ed. N. Eldredge & S. M. Stanley, pp. 4–22. New York: Springer Verlag.

Novacek, M. (1985). Cranial evidence for rodent affinities. In *Evolutionary Relationships among Rodents: a Multidisciplinary Analysis*, ed. W. P. Luckett & J.-L. Hartenberger, pp. 59–81. New York: Plenum Publishing Corporation.

Patterson, B. (1965). The fossil elephant shrews (Family Macroscelididae). *Bulletin of the Museum of Comparative Zoology, Harvard University*, **133**(6), 295–335.

Patterson, B. (1975). The fossil aardvarks (Mammalia: Tubulidentata). *Bulletin of the Museum of Comparative Zoology, Harvard University*, **147**(5), 185–237.

Penny, D., Foulds, L. R. & Hendy, M. D. (1982). Testing the theory of evolution by comparing phylogenetic trees constructed from five different protein sequences. *Nature, London*, **297,** 197–200.

Penny, D. & Hendy, M. D. (1985a). The use of tree comparison metrics. *Systematic Zoology*, **34**(1), 75–82.

Penny, D. & Hendy, M. D. (1985b). Testing methods of evolutionary tree construction. *Cladistics*, **1**(3), 266–78.

Prass, W. A., Berkley, D. S. & Romero-Herrera, A. E. (1983). Chicken cardiac myoglobin revisited. *Biochimica et biophysica Acta*, **742,** 677–80.

Presley, R. (1980). The braincase in Recent and Mesozoic therapsids. *Mémoires de la Societé géologique de France* (new series), **139,** 159–62.

Presley, R. (1981). Alisphenoid equivalents in placentals, marsupials, monotremes and fossils. *Nature, London*, **294,** 668–70.

Presley, R. & Steel, F. L. D. (1976). On the homology of the alisphenoid. *Journal of Anatomy*, **121**(3), 441–59.

Reed, C. A. (1960). Polyphyletic or monophyletic ancestry of mammals, or: what is a class? *Evolution*, **14,** 314–22.

Romer, A. S. (1966). *Vertebrate Paleontology*, 3rd edn. University of Chicago Press.

Romero-Herrera, A. E., Lehmann, H., Castillo, O., Joysey, K. A. & Friday, A. E. (1976). Myoglobin of the orangutan as a phylogenetic enigma. *Nature, London*, **261,** 162–4.

Romero-Herrera, A. E., Lehmann, H., Joysey, K. A. & Friday, A. E. (1978). On the evolution of myoglobin. *Philosophical Transactions of the Royal Society of London, Biological Sciences*, **283**(995), 61–163.

Rose, K. D. (1978). A new Paleocene epoicotheriid (Mammalia), with comments on the Palaeanodonta. *Journal of Paleontology*, **52,** 658–74.

Rose, K. D. (1979). A new Paleocene palaeanodont and the origin of the Metacheiromyidae (Mammalia). *Breviora*, **455,** 1–14.

Rose, K. D. (1982). Anterior dentition of the early Eocene plagiomenid dermopteran *Worlandia*. *Journal of Mammalogy*, **63**(1), 179–83.

Schaeffer, B. (1947). Notes on the origin and function of the artiodactyl tarsus. *American Museum Novitates*, **1356,** 1–24.

Shoshani, J., Goodman, M., Czelusniak, J. & Braunitzer, G. (1985). A phylogeny of Rodentia and other eutherian orders: parsimony analysis utilizing amino acid sequences of alpha and beta hemoglobin chains. In *Evolutionary Relationships among Rodents: a Multidisciplinary Analysis*, ed. W. P. Luckett & J.-L. Hartenberger, pp. 191–210. New York: Plenum Publishing Corporation.

Sibley, C. G. & Ahlquist, J. E. (1984). The phylogeny of the hominoid primates as indicated by DNA–DNA hybridization. *Journal of Molecular Evolution*, **20,** 2–15.

Simpson, G. G. (1945). The principles of classification and a classification of mammals. *Bulletin of the American Museum of Natural History*, **85,** i–xvi, 1–350.

Simpson, G. G. (1959). Mesozoic mammals and the polyphyletic origin of mammals. *Evolution*, **13**(3), 405–14.

Simpson, G. G. (1960). Diagnosis of the classes Reptilia and Mammalia. *Evolution*, **14**(3), 388–92.

Sonntag, C. F. (1925). A monograph of *Orycteropus afer*. I. Anatomy except the nervous system, skin, and skeleton. *Proceedings of the Zoological Society of London*, **1925**, 331–437.

Stapel, S. O., Leunissen, J. A. M., Versteeg, M., Wattel, J. & de Jong, W. W. (1984). Ratites as oldest offshoot of avian stem – evidence from alpha crystallin A sequences. *Nature, London,* **311,** 257–9.

Storch, G. (1981). *Eurotamandua joresi*, ein Myrmecophagide aus dem Eozän der 'Grube Messel' bei Darmstadt (Mammalia, Xenarthra). *Senckenbergiana Lethaea*, **61**(3/6), 247–89.

Sues, H.-D. (1985). The relationships of the Tritylodontidae (Synapsida). *Zoological Journal of the Linnean Society*, **85,** 205–17.

Sulimski, A. (1969). Paleocene genus *Pseudictops* Matthew, Granger and Simpson, 1929 (Mammalia) and its revision. *Palaeontologica Polonica*, (1968), **19,** 101–29.

Szalay, F. S. (1977). Phylogenetic relationships and a classification of the eutherian Mammalia. In *Major Patterns in Vertebrate Evolution*, ed. M. K. Hecht, B. M. Hecht & P. C. Goody, pp. 315–74. New York: Plenum Press.

Szalay, F. S. (1985). Rodent and lagomorph morphotype adaptations, origins, and relationships: some postcranial attributes analysed. In *Evolutionary Relationships among Rodents: a Multidisciplinary Analysis,* ed. W. P. Luckett & J.-L. Hartenberger, pp. 83–132. New York: Plenum Publishing Corporation.

Szalay, F. S. & Decker, R. L. (1974). Origins, evolution, and function of the tarsus in Late Cretaceous Eutheria and Paleocene primates. In *Primate Locomotion*, ed. F. A. Jenkins, Jr, pp. 223–59. New York & London: Academic Press.

Templeton, A. R. (1985). The phylogeny of the hominoid primates: a statistical analysis of the DNA–DNA hybridization data. *Molecular Biology and Evolution*, **2**(5), 420–33.

Thewissen, J. G. M. (1985). Cephalic evidence for the affinities of Tubulidentata. *Mammalia*, **49**(2), 257–84.

Van Valen, L. (1960). Therapsids as mammals. *Evolution*, **14**(3), 304–13.

Van Valen, L. (1971). Adaptive zones and the orders of mammals. *Evolution*, **25**(2), 420–8.

Vötsch, W. & Anderer, F. A. (1975). The N-terminal amino acid sequence of sheep heart myoglobin. *Zeitschrift für Naturforschung*, **27b**, 157–9 [Not seen. Cited in Romero-Herrera *et al.*, 1978, p. 68].

Walter, L. R. (1985). The formation of secondary centers of ossification in kannemeyeriid dicynodonts. *Journal of Paleontology*, **59**(6), 1486–8.

Woodburne, M. O. & Tedford, R. H. (1975). The first Tertiary monotreme from Australia. *American Museum Novitates*, **2588**, 1–11.

APPENDIX 1

Hypothesized amino acid substitutions in myoglobin and in the alpha crystallin A chain

In order to simplify drafting of the cladograms presented in Figs 3.1 to 3.6, all amino acid substitutions and deletions hypothesized at any particular node have been assigned a single number, the significance of which is given here. Parallel results of substitutions are marked with an equals sign (=); back mutations with a minus sign (−); and double hits with two asterisks (**). These may occur in combinations.

Myoglobin

1. See Table 3.1 for the hypothetical sequence of tetrapod myoglobin at this node. Other nodes modify this sequence in the ways indicated.
2. 5 asp, 9 his, 13 ile=, 19 pro= (=with a shark, *Heterodontus*), 29 ile, 34 gln, 48 asn=, 57 ser=, 60 glu=, 74 thr (ACU), 78 gln, 81 asn−, 110 ile= (=with *Heterodontus*), 112 val−=, 115 ile=, 127 ser, 132 arg−, 144 ser=.
3. 8 his=, 15 ala= (=with *Heterodontus*), 22 ser=, 23 ala, 35 val, 52 ile**, 55 leu−, 56 arg=, 61 val, 66 thr=, 74 arg**=, 78 leu, 80 asn, 84 pro, 91 glu=, 108 cys=, 121 ser= (=with *Heterodontus*).
4. 5 gln**, 8 lys, 26 his, 35 glu−, 44 glu.
5. −1 met, 1 glu, 19 ser=, 20 lys, 21 val=, 23 glu, 33 leu, 49 met, 52 ala=, 60 lys, 63 gln, 66 asn**=, 69 phe, 74 asn−, 83 ala=, 84 glu, 85 val, 91 lys, 95 leu**.
6. 9 gln−, 35 asp=, 41 asp=, 52 gln, 60 asp−, 61 leu=, 74 pro (CCU), 81 his−=, 92 thr−, 110 val−=, 129 glu.
7. 5 glu, 9 lys, 23 ser=, 26 gln−, 32 met, 35 asn=, 48 his=, 52 leu, 57 asn**, 66 thr=, 74 arg**=, 86 ile=, 87 ala=, 95 asn=, 96 thr= (=with a teleost fish, *Cyprinus*), 101 ile−=, 108 cys=, 113 gly, 120 ser=, 145 arg, 149 leu=.
8. 8 gln−, 12 thr, 22 ala=, 29 leu−, 30 met, 40 leu=, 44 asp=, 48 gly**=, 52 pro=, 57 gly=, 66 glu, 74 gln** (CAA), 112 ile=, 120 ala=.
9. 19 ala−=, 21 ile=, 34 his=, 53 asn=, 71 gln, 74 lys, 81 gln=, 84 ser, 85 asp, 129 ala−=, 140 asp.

10. 3 asn, 13 met, 19 ser=, 27 ala= (=with *Heterodontus*), 35 ser★★=, 40 met, 47 arg, 53 glu, 54 asp=, 56 arg=, 61 met−, 66 val=, 78 lys−, 90 ser, 99 val=, 110 ala=, 112 met, 117 gln, 121 ser=, 122 asn=, 127 ala−.
11. 9 leu=, 40 leu=, 42 lys, 44 asp=, 48 his=, 61 leu=, 103 phe, 110 ala=, 113 his★★=, 115 leu, 116 gln★★, 117 ser★★.
12. 12 lys=, 35 thr★★=, 59 ala=, 66 gly, 81 gln=, 82 his, 100 ser=, 101 ile−=, 120 ser=, 132 gly★★=.
13. 19 gly, 48 gly★★=, 106 tyr.
14. 19 thr=, 21 ile=, 22 thr=, 27 asp=, 67 val=, 74 ser, 144 thr=.
15. 21 ile=, 66 val=, 109 asp=, 121 gly=, 149 leu=.
16. 51 ser=, 102 gln, 113 gln=, 115 ile=.
17. 13 ala=, 22 pro=, 66 ala−=, 81 asn=, 100 ser=, 109 glu−=, 132 gly★★=.
18. 13 ile=, 19 thr=, 21 val=, 22 gly★★=, 27 asp=, 66 ile★★, 120 ala=, 140 his, 149 phe−=.
19. 74 gly★★, 103 tyr−, 142 ile.
20. 8 glu= (=with *Cyprinus*), 12 lys=, 13 thr, 26 glu=, 27 phe, 30 val, 34 thr=, 53 gly, 64 gln, 86 ile=, 87 gln, 101 ile−=, 121 ala−=, 122 glu=.
21. 129 gly=.
22. 21 leu−=, 22 ala=, 66 asn★★=.
23. 19 thr=, 35 ser=, 57 gly=.
24. 13 ile=, 35 asn=, 41 asp=, 113 gln=, 120 ser=, 124 his★★, 127 thr=, 128 glu.
25. 117 asn=.
26. 51 ser=, 57 arg.
27. 8 his=, 54 asp=, 56 arg=, 62 arg=, 101 ile−=, 109 glu−=, 116 his=, 121 ala−=, 122 glu=, 152 his=.
28. 19 ala−=, 35 gly−.
29. 13 ile=, 22 val, 66 lys, 83 asp=, 127 thr=, 128 his, 147 arg.
30. 21 val=, 53 ala=, 57 gly−=, 82 gln, 112 ala★★, 124 ala=, 126 glu, 129 gly=.
31. 21 val=, 129 gly=, 132 asn.
32. 53 ala=, 101 ile−=, 116 his=.
33. 27 asp=, 109 glu−=, 118 arg, 121 ala−=, 140 lys=.
34. 1 val, 15 ala=, 28 ile=.
35. 4 glu=, 12 his, 45 arg, 66 val★★=, 74 ala=, 151 tyr=.
36. 35 ser=, 121 gly−=.
37. 21 ile−=, 35 his, 50 ser, 51 ser=, 132 ser=.
38. 5 ala=, 13 ile=, 129 ala−=.
39. 121 gly−=.
40. 8 his=.
41. 122 glu=.
42. 3 thr, 31 ser=.
43. 21 leu−=, 122 glu=, 152 his=.
44. 83 asp=.
45. 54 asp=, 74 ala=.
46. 28 ile=.
47. 83 glu.
48. 21 val−=.
49. 4 glu=, 144 thr=.
50. 13 ile=, 121 gly−=, 122 asp−, 129 ala−=.
51. 9 thr, 21 ile−=, 34 thr=, 66 thr=, 67 val★★=, 132 thr.
52. 109 glu−=, 132 ser=.

53. 142 met−=.
54. 113 gln=, 115 ile=.
55. 13 ala=, 34 thr=, 53 ala=, 86 val= (=with *Heterodontus*), 88 his, 91 glu=, 95 asn=, 99 val=, 109 asp−=, 116 his=, 117 ala★★, 121 ser=, 122 asn=, 145 gln=, 148 val.
56. 101 ile−= (dictated by local parsimony!).
57. 9 ala= (=with *Thunnus*, tuna), 124 ala=, 142 ala★★, 144 glu, 145 lys−, 152 his=.
58. 81 gln=.
59. 113 gln=.
60. 21 leu−=, 51 ser=.
61. 22 pro−=, 66 thr=, 85 gln.
62. 21 ile−=, 66 ala=, 110 val=, 132 gly=.
63. 35 asp=, 57 ser=, 87 ala★★=, 95 asn=, 116 lys, 120 ala=.
64. 35 asn=, 48 asn=, 57 gly=, 59 asp, 70 ser=, 142 met−=.
65. 5 ala=, 34 his★★=, 35 thr=, 74 ala=, 81 his−=, 86 ile=, 113 his−=, 116 his=, 129 ala−=, 145 gln=.
66. 86 ile=.

66A. 5 ala=, 27 asp=, 45 arg=, 66 thr=, 87 gln=, 115 ile=, 120 ser=.

66B. 70 ser=, 129 ala−=.

67. 66 val★★−=.
68. 34 thr=, 52 ala=.
69. 21 leu−=, 35 ala, 48 asn=, 86 ile=.
70. 12 lys=, 13 ile=, 27 asp=, 117 asn=, 120 ser=, 125 thr, 127 val.
71. 9 ser★★, 28 ile=, 52 pro★★=.
72. 81 his−=.
73. 109 gly.
74. 22 gly=, 66 thr★★=, 101 ile−=, 109 asp=, 112 val−=, 116 his=, 121 ala−=, 122 glu=, 129 ala−=, 132 lys★★−=.
75. 21 ile−=, 22 pro−=, 23 ser=, 51 ser=, 142 met−=.
76. 31 ser=, 60 glu=, 109 asp−=, 112 val−=, 117 lys★★, 132 lys★★−=.
77. 81 his−=.
78. 66 ala=.
79. 13 ile=.
80. 66 thr★★=, 106 leu=.
81. 112 ile−=, 114 ala.
82. 66 ala=, 81 his−=, 86 ile=, 110 ser, 113 gln=.
83. 106 leu=.
84. 66 val−=.
85. 140 lys=, 144 ser=, 145 asn−.
86. 110 cys.
87. 23 gly−.
88. 116 his=.
89. 22 ser=.

Alpha crystallin A chain

90. See Table 3.2 for hypothetical primitive tetrapod sequence at this node. Other nodes modify this sequence in the ways indicated.

91. 20 asn=, 23 asp**, 26 val, 31 met, 33 asp, 37 phe, 52 phe, 55 gly, 56 phe, 70 arg=, 89 ile=, 124 leu, 125 asn, 126 glu, 138 ile, 146 leu=, 147 met, 149 ser=, 150 leu=, 155 gly=.
92. 55 thr.
93. 17 leu=, 18 ile**, 39 leu, 40 phe, 58 glu=, 72 thr, 86 ser, 90 ile, 130 thr, (?146 val= and ?149 asn, *fide* Stapel *et al.*, 1984), 150 met=, 152 pro=.
94. 55 ser−=, 74 met**, 89 ile=.
95. 32 leu, 135 ser.
96. 69 glu, 148 ala, (*Dromaius* 111 ala; *Struthio* 101 asn=).
97. 122 ala, 147 pro=.
98. 16 ser, 17 leu=, 58 glu=, 146 ile=, 147 his, 156 asp, 158 ser, 169 leu**.
99. 70 arg=, 74 tyr=, 93 tyr, 148 thr=, 151 glu.
100. 149 asp, 152 ala=.
101. 3 val, 101 asn=, 129 leu, 149 gly**, 152 ala=, 158 ala, 168 ser.
102. 56 ala (alternative: no substitution), 122 thr, 123 ala, 147–149 delete, 150 val=, 152 pro=, 158 thr.
103. 146 ile=.
104. 7 gln=, 91 gly, 158 pro.
105. ?11 arg (see Stapel *et al.*, 1984), 56 val− (alternative: no substitution).
106. (additional substitution possible: 56 ala), 70 arg=, 133–142 delete, 146 leu=, 155 gly=.
107. 74 leu=.
108. 70 gln, 74 leu=, 142 cys=.
109. 90 gln=, 146 ile=.
110. 72 leu.
111. 84 glu, 149 ser=, 152 asp, 153 gly=, 170 val.
112. 19 his, 20 asn=, 126 lys, 167 ala, 172 or 173 asn.
113. 91 glu, 150 leu=, 153 gly=.
114. 3 ile−, 150 val=.
115. 90 gln=, 146 leu, 152 gly.
116. 4 ala, 147 pro=.
117. 7 gln=, 37 met, 61 val=, 91 gln.
118. 74 tyr=, 150 met=. See caption of Fig. 3.3
119. 51 pro, 52 val, 55 ser=.
120. 153 ser−, 74 tyr=. See caption of Fig. 3.3
121. 93 ser, 101 ser−=, 152 thr.
122. 92 gly.
123. 90 gln=.
124. 150 met−=.
125. 55 ser=, 147 thr.
126. 146 ile=.
127. 127 thr, 150 met−=.
128. 13 thr=.
129. 55 ser=, 61 val=.
130. 148 val, 150 leu=.
131. 61 val=, 146 val−=, 172 thr.
132. 13 thr=.
133. 90 leu−=.
134. 23 leu, 101 ser−=, 122 pro, 123 ser, 127 ala, 147 ala, 154 pro.
135. 90 gln=.

136. 101 ser−=.
137. 13 thr=.
138. 90 leu−=.
139. 13 pro=.
140. 61 val=.
141. 133 val.
142. 13 thr=, 91 asp−, 146 ile=, 148 thr=, 153 delete, 155 thr.
143. 162 ala.
144. 142 cys=, 168 thr−.

APPENDIX 2

Genetic code

For convenience, the genetic code for Crick's (1958) magic 20 biologically active amino acids is given here, together with the one-letter and three-letter abbreviations for these amino acids.

Alanine	(A, ala)	GCU	GCC	GCA	GCG		
Cysteine	(C, cys)	UGU	UGC				
Aspartic acid	(D, asp)	GAU	GAC				
Glutamic acid	(E, glu)	GAA	GAG				
Phenylalanine	(F, phe)	UUU	UUC				
Glycine	(G, gly)	GGU	GGC	GGA	GGG		
Histidine	(H, his)	CAU	CAC				
Isoleucine	(I, ile)	AUU	AUC	AUA			
Lysine	(K, lys)	AAA	AAG				
Leucine	(L, leu)	UUA	UUG	CUU	CUC	CUA	CUG
Methionine	(M, met)	AUG					
Asparagine	(N, asn)	AAU	AAC				
Proline	(P, pro)	CCU	CCC	CCA	CCG		
Glutamine	(Q, gln)	CAA	CAG				
Arginine	(R, arg)	CGU	CGG	CGA	CGC	AGA	AGG
Serine	(S, ser)	UCU	UCC	UCA	UCG	AGU	AGC
Threonine	(T, thr)	ACU	ACC	ACA	ACG		
Valine	(V, val)	GUU	GUC	GUA	GUG		
Tyrosine	(Y, tyr)	UAU	UAC				
Tryptophan	(W, trp)	UGG					

APPENDIX 3

Characters hypothesized at nodes in Fig. 3.7

Characters are mainly from Kemp (1983), but winnowed after consideration of Sues (1985), Presley (1980, 1981), Presley & Steel (1976).

1. Enlarge dentary bone relative to the postdentary bones.
2. Adductor musculature invades lateral surface of coronoid process of dentary bone, as shown by adductor fossa.
3. Differentiate postcanine teeth into simpler anterior ones and more complex posterior ones.
4. Form anterior lamina and lateral process of periotic bone. (Not same lamina as in mammals, character 58?)
5. Form secondary palate in mammal-like fashion (in contrast to certain other therapsids that form a secondary palate in other ways; see also Sues, 1985, p. 206).
6. Differentiate vertebral column into a thoracic region with a rib cage and a lumbar region with reduced, immoveable ribs.
7. Spinatus musculature invades scapula.
8. Reduce coracoid bones.
9. Ilium extends forward; pubis reduced.
10. Dentary bone enlarges further, with ventral angular process and posteriorly directed articular process.
11. Reduce postdentary bones (including the reflected lamina of the angular).
12. Squamosal bone forms glenoid fossa for secondary articulation with lower jaw.
13. Postcanine teeth accurately occlude.
14. Reduce rate of tooth replacement.
15. Posterior part of infraorbital canal gives off three branches that go directly to external surface of skull.
16. Reduce coracoid and procoracoid bones further.
17. Expand palatine bone on palate.
18. Lose expanded ribs.
19. Anterior lamina (of periotic bone) is larger than in any known 'cynodont' and partially surrounds the trigeminal foramen. (But is it the same anterior lamina as in mammals? See Sues, 1985, p. 208.)
20. Reduce pila antotica (subject to variation: Sues, 1985, p. 208).
21. In medial view, the internal auditory meatus is walled, with separate foramina for the vestibular and cochlear branches of the auditory nerve.
22. Quadrate bone with dorsal extension.
23. Articulation of quadrate bone with articular bone has a ventral process.
24. Reduce squamosal bone dorsoventrally (paralleled by the advanced 'cynodont' *Exaeretodon*; Sues, 1985, p. 210). Expand parietal bone posteriorly.
25. Posterior palatine foramen forms.
26. Lose prefrontal bone (paralleled by the advanced 'cynodont' *Therioherpeton*; Sues, 1985, p. 207).
27. Lose postorbital bone (paralleled by the advanced 'cynodont' *Therioherpeton*; Sues, 1985, p. 207).
28. Vertebral centra platycoelous.
29. Beginning of differentiation of thoracic and lumbar vertebrae.
30. Neural arch of atlas vertebra is relatively shorter from front to back.
31. Odontoid process of axis vertebra is pronounced.
32. Neural spines of thoracic vertebrae low and extend horizontally backwards. Reduce transverse process.
33. Sternebrae ossify (not known certainly in morganucodontids).
34. Procoracoid excluded from glenoid fossa of scapula (paralleled in *Exaeretodon*; Sues, 1985, p. 211).
35. Acromion better developed and more everted than in 'cynodonts'.

36. Scapular part of glenoid fossa faces laterad rather than ventrad as in 'cynodonts'.
37. Ulna with very well developed olecranon process. (This character may possibly be at the wrong node because of a smaller olecranon in *Morganucodon*; Sues, 1985, p. 211.)
38. Lose posterior process of ilium.
39. Anterior process of ilium supports longitudinal lateral ridge.
40. Develop epipubic bones (doubtful in morganucodontids).
41. Femur develops trochanter minor on medial side. Lose trochanter internus of posterior side of femur of 'cynodonts' (or it converts to trochanter minor?).
42. Astragalus develops definite head for articulation with navicular bone.
43. Astragalus more superposed on calcaneum than in 'cynodonts'.
44. Calcaneum develops definite facet for articulation with cuboid bone.
45. Short, wide basicranial axis.
46. Quadrate bone contacts paroccipital process but not squamosal bone.
47. Lose squamosal/surangular contact.
48. Enlarge postcanine teeth for herbivory.
49. Prismatic enamel.
50. Cavum epiptericum develops a floor (except *Kayentatherium*: Sues, 1985, p. 209).
51. Reduced lateral descending flange of pterygoid bone.
52. Lose internarial bar of 'cynodonts' (composed of paired premaxillae: Sues, 1985, p. 207).
53. Expand lateral flange of periotic bone above a groove for the vena capitis lateralis.
54. Dentary bone replaces surangular bone in secondary jaw articulation.
55. Double-rooted postcanine teeth appear (Sues, 1985, p. 207).
56. Secondary jaw bones no longer associate with dentary.
57. Chewing action of molars involves medial movements of dentary bone.
58. Enlarge anterior lamina of periotic bone.
59. Develop reversed triangular cheek-tooth pattern.
60. Reduce alisphenoid (epipterygoid) to ala temporalis.
61. Reduce septomaxillary bone, whose exposure is now internal to nostril, not on face.
62. Expand neopallium.
63. Fuse neural arches, intercentrum, and centrum of atlas vertebra to form a complete bony ring.
64. Fuse cervical ribs to cervical vertebrae, forming transverse foramina.
65. Acetabulum with complete (non-notched) border.
66. Long bones develop epiphyses (other vertebrates also develop epiphyses; Walter, 1985).
67. Transform secondary jaw bones into auditory ossicles: quadrate = incus; articular = malleus; angular = tympanic.
68. Fuse anterior lamina to periotic bone.
69. Lose teeth.
70. Fuse anterior lamina to ala temporalis.
71. Interpose squamosal bone between periotic bone and anterior lamina (now part of alisphenoid bone).
72. Lose interclavicle bone.
73. Lose procoracoid bone.
74. Begin to rotate plane of tympanic ring from nearly horizontal to more nearly vertical.
75. Molars tribosphenic.
76. Auditory bulla forms from alisphenoid.

4 Avian phylogeny reconstructed from comparisons of the genetic material, DNA

Charles G. Sibley & Jon E. Ahlquist

Introduction

All organisms have an evolutionary history, because all organisms had ancestors. The reconstruction of this history, or phylogeny, has long been one of the most challenging problems in evolutionary biology. Because all life on Earth presumably had a single origin, and certainly had a single history, there is only one true phylogeny. The question is, how can it be reconstructed?

The pattern of phylogeny is like that of the branchings of a huge bushy tree; the origin of life is represented by a single root that gave rise to a short trunk that branched, and branched again and again, until the uppermost branches are single twigs. In the tree of life, each of the billions of living individual plants and animals is represented by a topmost twig of this vast, flat-topped tree. It is flat-topped because the vertical dimension of the tree is time, and the plane of the flat top is today. Each living human individual is represented by one of the topmost twigs, thus we can see only the other living individuals of plants and animals. All of the tree below the living twigs at the top represents the ancestors which, except for fossil remains, are invisible to us. Many branches were broken off short; they are the extinct lineages which have no descendants and may, or may not, be represented by fossils. Our problem is to reconstruct the *branching pattern* of the tree and, if possible, to *date* each branching event. If we can do these two things, we will have reconstructed the phylogeny of life on Earth.

We begin this task by clustering individuals into larger and larger groups, based on some hypothesis about what constitutes a valid basis for such clustering. Evolution provides that hypothesis: we will cluster on the basis of genealogical relationship, i.e. we will look for evidence of the degree of genetic relatedness. The first clusters will contain the members of the same human family, or litter of mice, clutch of eggs, seeds from the same plant, etc. Using the same concept, we will assemble these primary clusters into populations, and species. Up to this point we are dealing mostly with the living terminal twigs; we can examine the members of these groups directly and assess their relationships from many sources of evidence, including the ability to interbreed, anatomical similarities, behaviour, distribution, etc. The branching events between closely related living species occurred in the recent past and in most cases the evidence of their relationships is clear and we make few mistakes. However, as we attempt to descend the tree into the more and more distant past, the problem of clustering becomes increasingly difficult.

The evidence of ancient branchings is preserved in some form in their living descendants; the problem is to interpret that evidence correctly. Morphology is an obvious source of evidence of relationship because we can expect closely related organisms to be similar in structure. This is usually true, but morphological characters are prone to *convergent evolution*, i.e. unrelated organisms come to look alike because their bodies have become adapted to cope with the same environmental demands through natural selection. For example, swifts and swallows are superficially similar because both are specialized to feed on flying insects, but swifts are more closely related to hummingbirds and nightjars than to swallows, and swallows are more closely related to the other songbirds (warblers, thrushes, sparrows, etc.) than to swifts. In early classifications swifts and swallows were placed together, and the debate about their relationships extended into the late 1800s, long after Charles Darwin had provided the basis for understanding convergent evolution.

Convergence can be far more subtle than the swift–swallow example, and it is a hazard in all comparisons of anatomical characters, whether in living or fossil organisms. The fossil record for most groups is fragmentary, and for some nearly non-existent. Partial phylogenies for some groups have been reconstructed from fossils, and it is possible to date most fossils with reasonable accuracy. However, a dated fossil records the time the individual organism died, not the time that its lineage diverged from that of its next nearest relatives.

Morphological comparisons and fossils have provided us with the evidence to reconstruct many portions of the tree of life. The avian fossil record is less complete than it is for other groups of vertebrates because bird bones are fragile and most birds tend to die in circumstances not conducive to fossilization. Even distantly related birds tend to be similar in many aspects of morphology because their structure has been constrained by the requirements of flight. Nonetheless, we can be certain that all parrots are more closely related to one another than to hawks, or pigeons, or ducks. Similarly, morphological similarities among hawks, pigeons, ducks, and many other groups of birds tell us that the members of such groups have descended from a common ancestor; i.e. they are *monophyletic* clusters. However, when we try to assess the evidence for the older branches that must link the more recent branches together, we encounter great difficulties. Are parrots more closely related to hawks than to pigeons? Are ducks more closely related to pheasants than to grebes? Relevant fossil evidence is lacking and anatomical comparisons reveal little or produce conflicting opinions from different studies of the same groups. The branchings occurred long ago, and adaptive changes, extinctions and convergence may erase the phylogenetic trail or provide false clues. Clearly, a method that is independent of fossils and morphology, not prone to convergence, and that can measure the genealogical distances between living species should improve the reconstruction of avian phylogeny. Since we are trying to measure degrees of genetic relationship, it makes sense to look for help in the genetic material itself.

DNA–DNA hybridization

The technique

During the past 30 years, molecular biologists have developed techniques that make it possible to decipher the information encoded in the genes – the DNA – of organisms. At the genetic level, information is encoded in sequences of units analogous to the letters of the alphabet that make up the sentences of a book. It has become clear that evolution at the genetic level involves changes in one or more 'letters' of the genetic message. If a method for comparing the DNAs of different species can be developed, it should be possible to assess the degrees of genealogical relationship among living organisms. DNA–DNA hybridization is such a method.

The nucleus of each gamete (sperm or egg) of a 'higher' organism contains one complete set of chromosomes, the genome, composed of deoxyribonucleic acid, or DNA. The nuclei of body cells contain two genomes. Thus, the genome is the haploid number of chromosomes; body cells contain the diploid number. The 23 chromosomes in a human genome contain several thousand genes, which are composed of about 2.5 thousand million (= 2.5 billion) coding units, or nucleotides. An avian genome contains about 2 billion nucleotides. With the technique of DNA–DNA hybridization it is possible to obtain a measure of the degree of genetic similarity between the genomes of different species, and thus to infer the genealogical distances between living species that represent different lineages that branched from one another millions of years ago. From these data we can reconstruct the branching pattern of the phylogeny and, by calibrating the DNA distance measures against time, determine the approximate dates of the branching events.

The structure of DNA is familiar to all biologists; its essential features for

DNA–DNA hybridization are its double-stranded structure, and the fact that the two strands are held together by hydrogen bonds. The strands are composed of complementary sequences of four kinds of chemical subunits, the nucleotides. Each nucleotide is composed of a five-carbon sugar, a phosphate group and a 'base'. The nucleotides differ from one another only in the structures of the four bases: adenine (A), thymine (T), cytosine (C) and guanine (G). The bases occur as complementary pairs: an A in one strand pairs only with a T in the other strand, a C pairs only with a G. A–T pairs are held together by two hydrogen bonds; C–G pairs by three. Genetic information is encoded in the sequence of the bases.

If double-stranded DNA in solution is heated to boiling, the hydrogen bonds between the A–T and C–G base pairs rupture, or melt, and the double-stranded structure is converted into single strands. The hydrogen bonds between base pairs are the weakest bonds in the DNA molecule, thus they can be melted without damaging the rest of the structure.

As a melted sample of DNA cools, the single strands, by chance, will collide with one another. If two colliding strands have complementary base sequences, they will 'recognize' one another and quickly reassociate into the double-stranded form as the hydrogen bonds between base pairs are re-established. If the temperature of the solution is above about 50 °C, the original sequence will reform. For example, if a sample of DNA is melted in a salt solution of moderate concentration (e.g. 0.12 M sodium phosphate), then allowed to cool to 60 °C and incubated at that temperature for several hours, most of the single strands will find complementary partners and the accuracy of base pairing will be virtually perfect. Given enough time, all of the single strands will find partners. (See the Appendix for a technical description of DNA–DNA hybridization.)

This reaction can also be carried out between the DNAs from different species. If the single-stranded DNAs from two species are combined and incubated at 60 °C for at least 100 hours, 'hybrid' double-stranded molecules will form between homologous base sequences, i.e. between sequences that are complementary because they are the descendants of the same ancestral base sequence that was present in the most recent common ancestor of the two species. Because the two species have been evolving since the divergence of their lineages, their homologous base sequences will contain mismatched bases at various positions. For example, a T in one strand of the hybrid molecule may be opposite a C, and no hydrogen bonds will form between them. Since the melting temperature of duplexes is proportional to the number of hydrogen bonds between the two strands, such mismatches will cause the hybrid DNAs to melt at a temperature lower than that required to melt the perfectly base-paired duplexes formed between the strands from the same species. Thus, by comparing the melting behaviour of homoduplexes (from the same species) with that of heteroduplexes (hybrids between different species), it is possible to obtain a measurement that is proportional to the number of mismatched bases in the two species forming the DNA–DNA hybrid.

The DNA molecular evolutionary clock

Mismatched bases result from the genetic changes (mutations) that have been fixed since the two species forming a DNA–DNA hybrid diverged from a common ancestor. Evolution is a time-related process, therefore the number of mismatched bases is *proportional* to the time that has passed since the two lineages diverged. Thus, the melting temperature of a DNA–DNA hybrid is an indirect measure of the elapsed time since the two species last shared a common ancestor. To calibrate the DNA 'clock', we can use dated geological events that caused branching events in the past. For example, as the protocontinent of Gondwanaland broke apart in the Cretaceous Period, the Atlantic Ocean gradually opened between Africa and South America. The gap became an impassable barrier for flightless animals about 80 million years before the present (MyrBP). We may assume that the ancestor of the living Ostrich of Africa and its closest living relatives, the two species of rheas of South America, was divided by the opening of the Atlantic into two species that evolved into the Ostrich and rheas during the past 80 Myr. By comparing the DNAs of an Ostrich and a Common Rhea, we have been able to calibrate the DNA distance measurements in terms of absolute time. Several similar comparisons have also been used, and all give approximately the same calibration, namely, a difference in the median melting temperature of 1 °C between two DNAs indicates that they diverged about 4 to 5 million years ago. The median melting temperature for Ostrich × Rhea DNA–DNA hybrids is about 17 °C below the median melting temperature of an Ostrich × Ostrich, or a Rhea × Rhea DNA hybrid. The difference in median melting temperature between that of the homoduplex hybrid (e.g. Ostrich × Ostrich) and that of the heteroduplex hybrid (Ostrich × Rhea) is the delta $T_{50}H$, in this case 17. The $T_{50}H$ is the temperature at which 50 per cent of the hybrid duplexes have melted into single strands, and 50 per cent remain double-stranded, i.e. the median melting temperature.

For the Ostrich × Rhea comparison, a calibration constant is obtained by dividing 80 Myr by 17, which is 4.7; thus, for each degree centigrade difference in the median melting temperature of a DNA–DNA hybrid, approximately 4.7 Myr have elapsed since the two species diverged from their most recent common ancestor. We have seven such calibrating events which give an average constant of 4.5 Myr per delta $T_{50}H$ 1.0. This constant is tentative and subject to correction, but we can use it to provide an approximate dating for divergence events.

From numerous comparisons of the DNAs of birds, we have found that the *average* rate of DNA evolution tends to be uniform in many groups, but is modified by the effects of generation time. That is, those groups of birds having delayed maturity (i.e. they do not breed until 2 or more years old) show a slower rate of genomic evolution compared to species which breed at the age of 1 year. This results in shorter branch lengths for those with delayed maturity, and makes such groups appear to be closely related to the members of other groups which, among themselves, may be quite divergent. Fortunately, most passerine birds and some non-passerines begin to breed at the age of 1 year, and the age structure for many species is similar. This makes it possible to use their rate of genomic evolution as a baseline against which the rates for birds with delayed maturity appear to be slowed down, and a correction can be made by using the relative rate test. In this study, the branches in Figs 4.1 and 4.2 have been corrected for shortened branch lengths in the appropriate groups, and the 1:4.5 ratio for the time calibration is thereby corrected.

There is a common perception that comparisons of the actual base sequences of one, or a few, genes should yield a better estimate of phylogenetic relationships than the comparisons of the average (or median) differences between entire genomes provided by DNA–DNA hybridization. However, there is evidence that the comparison between the sequences of several genes will give the same answer as the comparison of median sequence divergence obtained by DNA–DNA hybridization. DNA sequencing is still a relatively slow process, but with DNA hybridization it is possible to make a hundred, or more, comparisons per week. The choice between the two methods may depend more on their rates of data production, than on differences in their answers to phylogenetic questions.

As noted above, morphological similarities may be the result of convergent evolution, as in the swift–swallow example. Fortunately, DNA–DNA hybridization data are immune to convergence, because the conditions of the experiments preclude the formation of heteroduplexes between non-homologous sequences. To form a stable duplex DNA molecule at 60 °C, 80 per cent of the bases in the two strands must be correctly paired, and *only homologous sequences* have this degree of complementarity. This solves the problem of homology and thereby eliminates the possibility of convergence.

During the past 10 years we have made more than 25 000 DNA–DNA hybrid comparisons among birds. The DNAs of about 1700 of the approximately 9000 living species, representing all but 3 of the 171 traditional 'families' of birds, have been compared. The following examples illustrate the ability of DNA hybridization to solve problems of phylogenetic reconstruction.

Results

The ratite birds (Fig. 4.1)

The living ratite birds are the Ostrich of Africa, the two rheas of South America, the Emu and three species of cassowaries of Australia and New Guinea, and the three species of kiwis of New Zealand. The next nearest relatives of the ratites are the tinamous, a group of about 47 species of chicken-like birds that occur in the New World from Mexico to Patagonia. The ratites are flightless but the ground-dwelling tinamous fly well.

Before the geological evidence of continental drift became convincing, most ornithologists thought that the ratites of each separate area had evolved

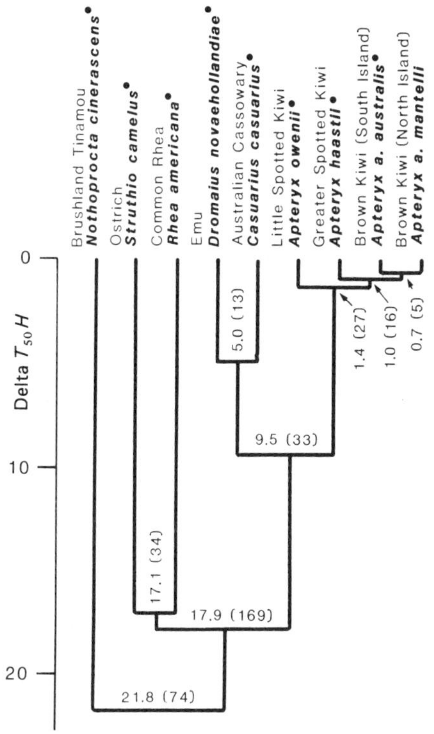

Fig. 4.1 Phylogeny of the ratites and tinamous as indicated by DNA–DNA hybridization. In this and the following figures, the numbers at each branch are the average delta $T_{50}H$ and, in parentheses, the number of DNA hybrids that were averaged; the spots after species names indicate radio-labelled or 'tracer' DNAs (see Appendix for details).

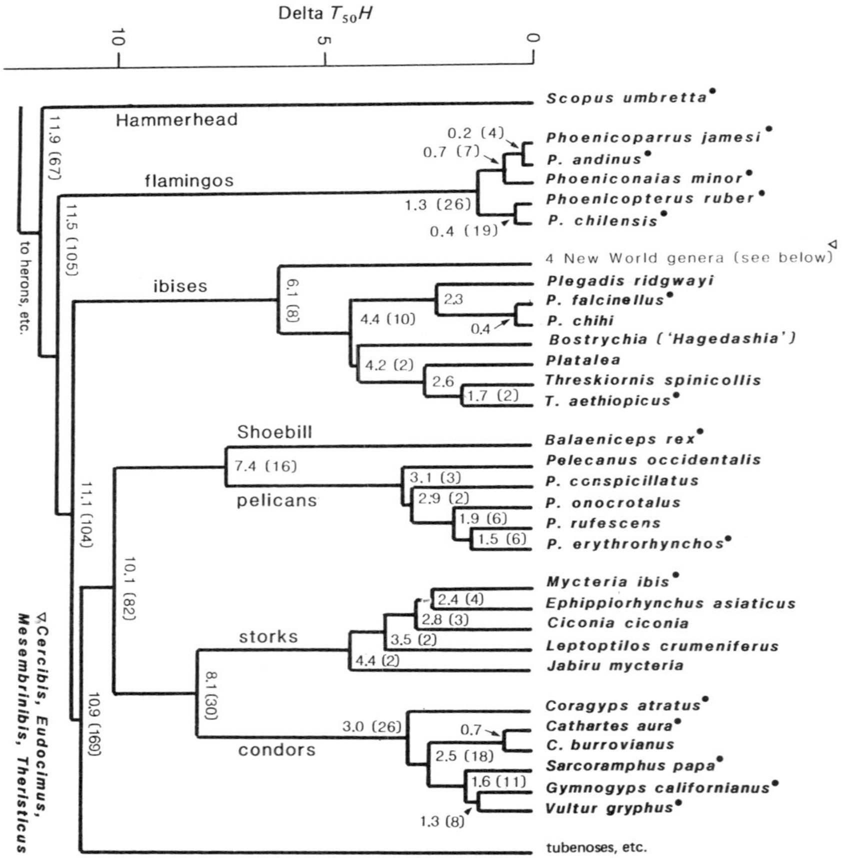

Fig. 4.2 Phylogeny of the Hammerhead, flamingos, ibises, Shoebill, pelicans, storks, and New World vultures, or condors. The four New World genera of ibises indicated by the triangle, and listed at the bottom of the Figure, are equidistant from the two radio-labelled species, *Plegadis falcinellus* and *Threskiornis aethiopicus*, but the relationships among the four genera can be determined only by labelling one or more of their DNAs and making comparisons among them.

independently from unrelated ancestors. This was logical because the vast areas of open ocean between them seemed to preclude any other explanation. There was some morphological and protein evidence of common ancestry, but it was often ignored. We (Sibley & Ahlquist, 1981) compared the DNAs of the ratite species with one another and with those of many other groups of birds. It was apparent that they had indeed shared a common ancestor, and that they are one another's closest living relatives. One of the surprises was that the small, chicken-sized kiwis are the closest relatives of the much larger Emu and cassowaries, and that the divergence between them occurred about 40–45 MyrBP. At that time, the Tasman Sea between Australia and New Zealand was already wide; it had begun to open 40 Myr earlier, in the Cretaceous. How then, did the flightless ancestor of the kiwis get from Australia to New Zealand? The answer was found in the geology of the area. The sea floor of the northern Tasman Sea is littered with the now submerged remains of ancient volcanic islands and submarine ridges that were emergent in the past. As Australia drifted northward from its original position in contact with Antarctica, it created volcanic islands and island arcs as the Australian crustal plate collided with the Pacific plate. The same thing is happening today in many parts of the world where volcanos and earthquakes are common. The islands were like a pathway of stepping stones for the ancestor of the kiwis, which may have been a large, flightless bird like their close relatives, the extinct moas. The ancestor may have taken several millions of years to make the crossing, and may have had to swim only short distances as islands arose, were eroded away, and new ones developed. The ancestor of the endemic frogs (*Leiopelma*) of New Zealand may have been present on proto-New Zealand when the Tasman Sea opened, but why the marsupial mammals were unable to colonize New Zealand from Australia is a mystery.

Vultures and storks (Fig. 4.2)

The vultures of Africa and Eurasia are closely related to the hawks and eagles; they are, in effect, carrion-eating eagles. The vultures of the New World, such as the condors (*Vultur, Gymnogyps*), Turkey Vulture (*Cathartes*), Black Vulture (*Coragyps*), and King Vulture (*Sarcoramphus*), are similar in external appearance to the Old World vultures, and the two groups have usually been classified together in the same order of birds, the Falconiformes. However, as long ago as the 1870s, Alfred Garrod of the Zoological Society of London pointed out numerous differences between the two groups of vultures, and also noted that the New World vultures share many morphological characters with the storks. The evidence of a stork–condor relationship was debated for a century, but most classifications continued to ally the two groups of vultures, although it was widely recognized that they are not closely related. In 1967, David Ligon extended and confirmed Garrod's observations; again the evidence was ignored by avian taxonomists. DNA comparisons have shown that those who advocated a close relationship between the New World vultures and the storks were correct. The superficial morphological similarities between the Old and New World vultures are due to convergence, not recent common ancestry (Sibley & Ahlquist, 1985*b*).

The flamingos (Fig. 4.2)

Another long-standing debate concerns the relationships of the flamingos. In their webbed feet, duck-like bill structures, and precocious downy young, flamingos are similar to the waterfowl (ducks, geese, swans), but their long necks and legs, and other characters, suggest that they shared an ancestry with the ibises and storks. Further to complicate the picture, certain fossils have been interpreted as evidence that flamingos are related to shorebirds, specifically to the avocets and stilts (Olson & Feduccia, 1980). DNA comparisons show that the flamingo lineage branched most recently (about 50 MyrBP) from a common ancestry with the ibises, storks, New World vultures, and pelicans, and that the divergence between these groups and either the waterfowl or the shorebirds was much earlier. Therefore, morphological similarities between flamingos and waterfowl or shorebirds must be due to convergence, not recent common ancestry (Sibley & Ahlquist, 1985*b*).

The Shoebill, pelicans, and the totipalmate birds (Fig. 4.2)

The Shoebill (*Balaeniceps*) is a large, stork-like bird that lives in the papyrus swamps of east Africa. Its enormous bill is adapted for feeding on lungfishes and other swamp creatures. The relationships of the Shoebill have been thought to be with the storks or herons, but nearly 30 years ago an anatomical study of the head region concluded that the Shoebill is a close relative of the pelicans (Cottam, 1957). This was dismissed by most taxonomists as an example of convergence, but DNA comparisons have shown that the Shoebill and the pelicans are one another's closest living relatives (Sibley & Ahlquist, 1985*b*).

The same DNA comparisons that revealed the Shoebill–pelican alliance also produced an even more startling result. The group of birds that have all four toes webbed, the 'totipalmate' birds, includes the pelicans (*Pelecanus*), cormorants (*Phalacrocorax*), anhingas (*Anhinga*), gannets and boobies (*Sula, Morus*), frigatebirds (*Fregata*), and tropicbirds (*Phaethon*). All but the tropicbirds have an obvious gular pouch between the branches of the lower jaw. These are the only birds with the foot webs connecting all four toes and, with the added evidence of the gular pouch, there has *never* been any doubt expressed about their monophyletic relationship. However, the DNA comparisons reveal that all of the 'Pelecaniformes' did not share a most recent common ancestor; instead, the pelicans are closer to the Shoebill, storks, ibises, condors, and flamingos (Fig. 4.2) than to any of the other totipalmate birds. The cormorants, anhingas and boobies are related to one another, and the frigatebirds are related to the tube-nosed swimmers (Procellariidae; albatrosses, petrels), the penguins (Spheniscidae) and the loons, or divers (Gaviidae). The tropicbirds seem to be an isolated group for which our studies have found no close relatives.

These discoveries represent radical departures from the traditional ideas about the relationships of these groups and they will not be readily accepted by many avian systematists, but we predict that morphological characters will be found that support the DNA data.

The passerine birds

More than half of the 9000+ species of living birds are members of the order Passeriformes, the so-called 'perching birds', including most of the small species such as warblers, flycatchers, thrushes, sparrows, crows, jays, starlings, swallows, larks, etc. (see Figs 4.3 to 4.8). The synopsis of passerine classification in Table 4.1 is based on the phylogeny developed from DNA comparisons.

The New World suboscine passerines, suborder Tyranni (Fig. 4.3)

Most of the passerine birds of South America are 'suboscines', of the suborder Tyranni, characterized by the structure of the vocal apparatus (syrinx) and several other morphological characters that separate them from the 'oscines', of the suborder Passeri, the songbirds. The New World suboscines include such groups as the tyrant flycatchers (Tyranninae), cotingas (Cotinginae), manakins (Piprinae), ovenbirds (Furnariidae), and antbirds (Formicariidae, Thamnophilidae). It is quite certain that these groups, and their relatives (see Fig. 4.3), evolved in South America during the long period in the Tertiary when South America was an island, isolated from other continents after the break-up of Gondwanaland until it was rejoined to North America about 3 MyrBP.

The New World suboscines provide examples of original discoveries based

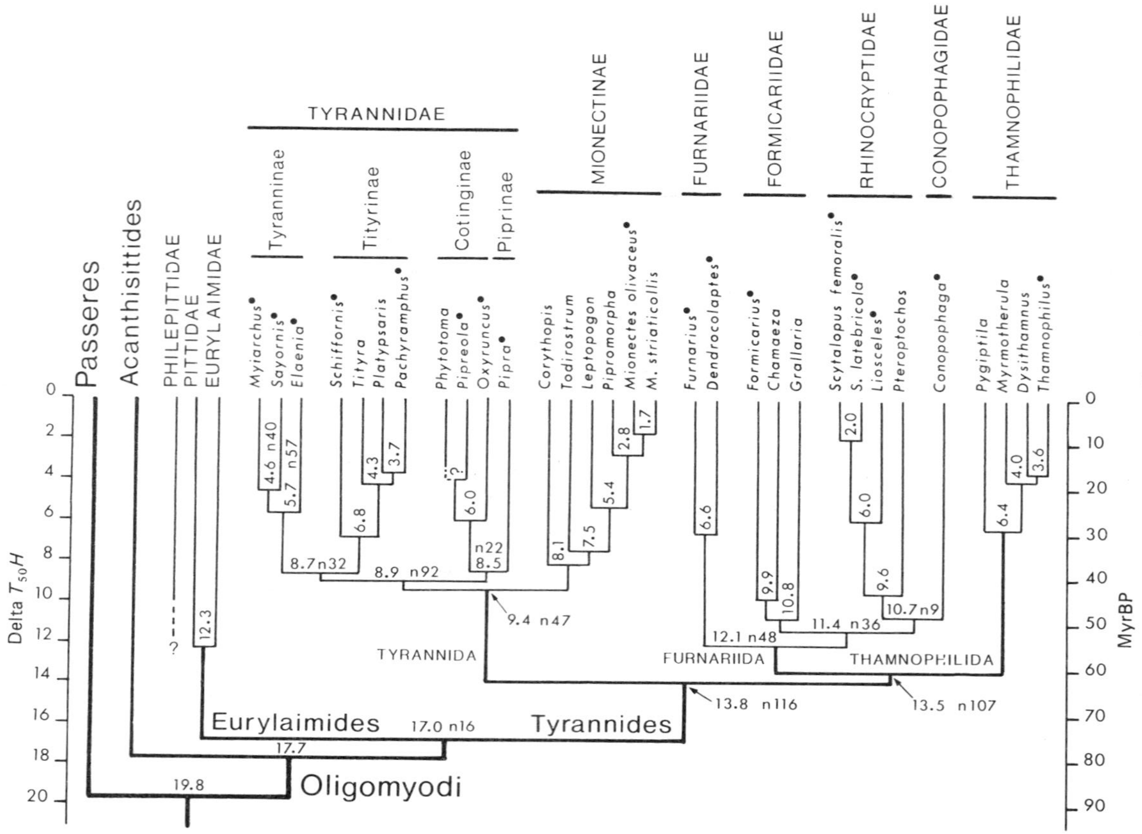

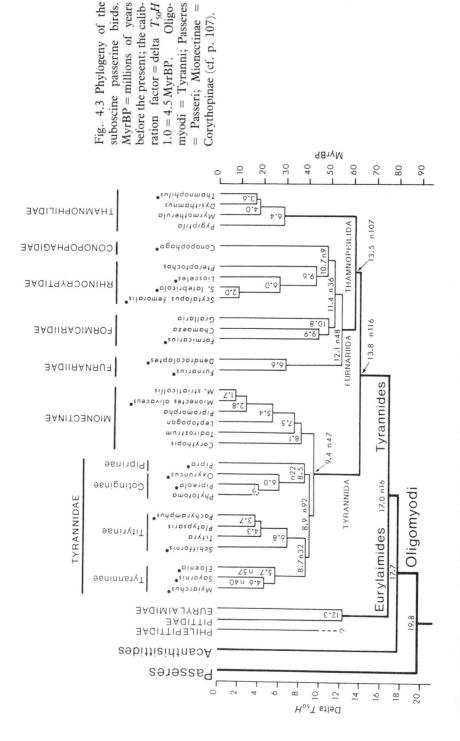

Fig. 4.3 Phylogeny of the suboscine passerine birds. MyrBP = millions of years before the present; the calibration factor = delta $T_{50}H$ 1.0 = 4.5 MyrBP. Oligomyodi = Tyranni; Passeres = Passeri; Mionectinae = Corythopinae (cf. p. 107).

Table 4.1. *Outline classification of passerine birds developed from DNA comparisons*

Some group names have been changed from those used in our recent publications; for example, the parvorders end in -ida, thus the Corvi have become the Corvida, and the Muscicapae = Passerida. Some changes were dictated by the International Rules of Zoological Nomenclature; thus the Turdoidea = Muscicapoidea, and Fringilloidea = Passeroidea.

- Order PASSERIFORMES
  - Suborder TYRANNI: suboscines (Fig. 4.3)
    - Infraorder ACANTHISITTIDES: New Zealand wrens
    - Infraorder EURYLAIMIDES: Old World Suboscines = pittas, broadbills, asities
    - Infraorder TYRANNIDES: New World suboscines
      - Parvorder TYRANNIDA: tyrants, becards, cotingas, manakins, mionectines
      - Parvorder THAMNOPHILIDA: typical antbirds
      - Parvorder FURNARIIDA
        - Superfamily FURNARIOIDEA: ovenbirds, woodcreepers
        - Superfamily FORMICARIOIDEA: ground antbirds, tapaculos, gnateaters
  - Suborder PASSERI: oscines or songbirds
    - Parvorder CORVIDA: oscines of Australian origin
      - Superfamily MENUROIDEA: lyrebirds, bowerbirds, treecreepers (Fig. 4.4)
      - Superfamily MELIPHAGOIDEA: fairy-wrens, honeyeaters, thornbills, pardalotes, scrubwrens (Fig. 4.5)
      - Superfamily CORVOIDEA: Australian robins, Australian flycatchers, log-runners, quail-thrushes, Apostlebird, Australian Chough, whistlers, sittellas, shrike-tits, Crested Bellbird, crows, ravens, currawongs, birds of paradise, orioles, cuckoo-shrikes, fantails, drongos, monarchs, bush-shrikes, vangas, vireos, shrikes, fairy-bluebirds, leafbirds (Figs 4.6 and 4.7)
    - Parvorder PASSERIDA: Afro-Eurasian–American oscines (Fig. 4.4)
      - Superfamily MUSCICAPOIDEA: waxwings, dippers, Old World flycatchers, thrushes, starlings, mockingbirds, thrashers (Fig. 4.8)
      - Superfamily SYLVIOIDEA: nuthatches, creepers, wrens, gnatcatchers, tits, chickadees, swallows, kinglets, bulbuls, white-eyes, Old World warblers, babblers
      - Superfamily PASSEROIDEA: larks, sunbirds, sugarbirds, flowerpeckers, sparrows, waxbills, wagtails, accentors, weavers, finches, Hawaiian honeycreepers, buntings, wood warblers, cardinals, meadowlarks, troupials, tanagers (including Neotropical honeycreepers = 'coerebids')

---

on DNA hybridization, and of congruence between morphological characters and the DNA comparisons. The DNA data revealed the existence of two monophyletic groups that had not been recognized by morphological comparisons. One of these is the Mionectinae, composed of species that had always been included in the Tyranninae, the tyrant flycatchers. The mionectine lineage branched from the tyrants before the tyrants separated from the cotingas and manakins (Sibley & Ahlquist, 1985*c*).

Another discovery involved the antbirds, an assemblage of more than 200 species which had been placed together in the Formicariidae. However, Heimerdinger & Ames (1967) had discovered that some formicariids have two notches in the posterior margin of their sterna (breastbones), while others have four notches. The same two groups also differ in the musculature of the syrinx (Ames, 1971). When we compared the DNAs of these birds we found the same groupings and were able to show that the antbirds with a two-notched sternum, and the associated syringeal characters, had branched off from the four-notched group much earlier than the four-notched 'ground antbirds' had branched from the tapaculos (Rhinocryptidae) and gnateaters (Conopophagidae), which also have four-notched sterna. Thus, the sternal and syringeal characters delineated the same groups as the DNA comparisons. The DNA–DNA data provided the branching order and the approximate datings of divergence events, thus permitting the recognition of a new parvorder (Thamnophilida) for the two-notched antbirds. The four-notched groups are members of the parvorder Furnariida, and are placed together in the superfamily Formicarioidea.

The suborder Passeri: oscines, or songbirds (Figs 4.4 to 4.8)

About 4000 of the passeriform species are members of the suborder Passeri, the 'songbirds'. The classification of the Passeri into subgroups has been difficult because they are morphologically similar in many characters, differing principally in the structure of the beak, tongue, feet and other superficial characters, most of which are obviously adapted for different modes of feeding. Clusters of supposedly related species have been based on the shape of the bill; thus, several groups of nectar-feeding species have been placed either in the same family or in adjacent families in virtually all classifications since Linnaeus. Similarly, species with broad, flat bills that forage on flying insects have been placed together as 'flycatchers', those with conical, seed-cracking bills, are assigned to the 'finches' or 'sparrows', and small species that pick insects from the foliage have been assembled as 'warblers'. Other such groups containing morphologically similar species from different parts of the world include 'nuthatches', 'creepers', 'babblers' and 'thrushes'.

The Australian songbirds (oscines) (Figs 4.4 to 4.7)

The songbirds of Australia were discovered many years after European ornithologists had classified the birds of most of the rest of the world. The specimens from Australia seemed to fit well into the groups of warblers, flycatchers, creepers, etc., that had been defined on the basis of Eurasian, African and American types, and there was no obvious evidence to the contrary. Thus, in most of the classifications of the past century, the family 'Sylviidae' included the 'warblers' of Australia, Eurasia and Africa, and some of the passerines of the New World. The 'flycatchers' of Australia, Eurasia and Africa were placed in the 'Muscicapidae', and the 'creepers' of the world were assigned to the 'Certhiidae'.

When we compared the DNAs of Australian songbirds with their supposed close relatives from other continents, we found a quite different situation. Instead of having their closest relatives in Africa, Asia or Europe, the Australian warblers, flycatchers, nuthatches, creepers, etc., turned out to be more closely related to one another than to the morphologically similar species from other continents. We discovered that the 4000 species of Passeri consist of two major groups, one of which (parvorder Corvida) evolved in Australia and New Guinea, and the other (parvorder Passerida) in Africa, Eurasia and North America (Fig. 4.4). Convergent evolution had produced morphologically similar species in different areas, and taxonomists had assembled them into clusters containing members of both parvorders. We estimate that the divergence between the two parvorders occurred about 50–60 MyrBP (Fig. 4.4) (Sibley & Ahlquist, 1985*a*).

The explanation for this situation is probably as follows. In the Cretaceous, approximately 90 MyrBP, Australia broke away from Antarctica and began to drift northward. By the early Tertiary (about 60 MyrBP), Australia was far south of its present position, and probably isolated from all other large land masses. At about 55–60 MyrBP, the ancestral species of the Corvida arrived in Australia, probably from Asia. During the early to middle Tertiary (50–30 MyrBP), the descendants of the early members of the Corvida became divided into many species which adapted to the environments of Australia and produced a great variety of endemic specialized forms, including the warbler-like thornbills (*Acanthiza*, Fig. 4.5), the nuthatch-like sittellas (*Daphoenositta*, Fig. 4.6), the creeper-like treecreepers (*Climacteris*, Fig. 4.4), the flycatcher-like monarchs (*Monarcha*, Fig. 4.6), and other groups that are the ecological and structural counterparts of members of the Passerida. Many of the convergences are so subtle that it is doubtful if the true relationships of these groups could have been resolved from anatomical comparisons. The parallel with the marsupials is obvious, but the marsupials have a marsupium and a unique

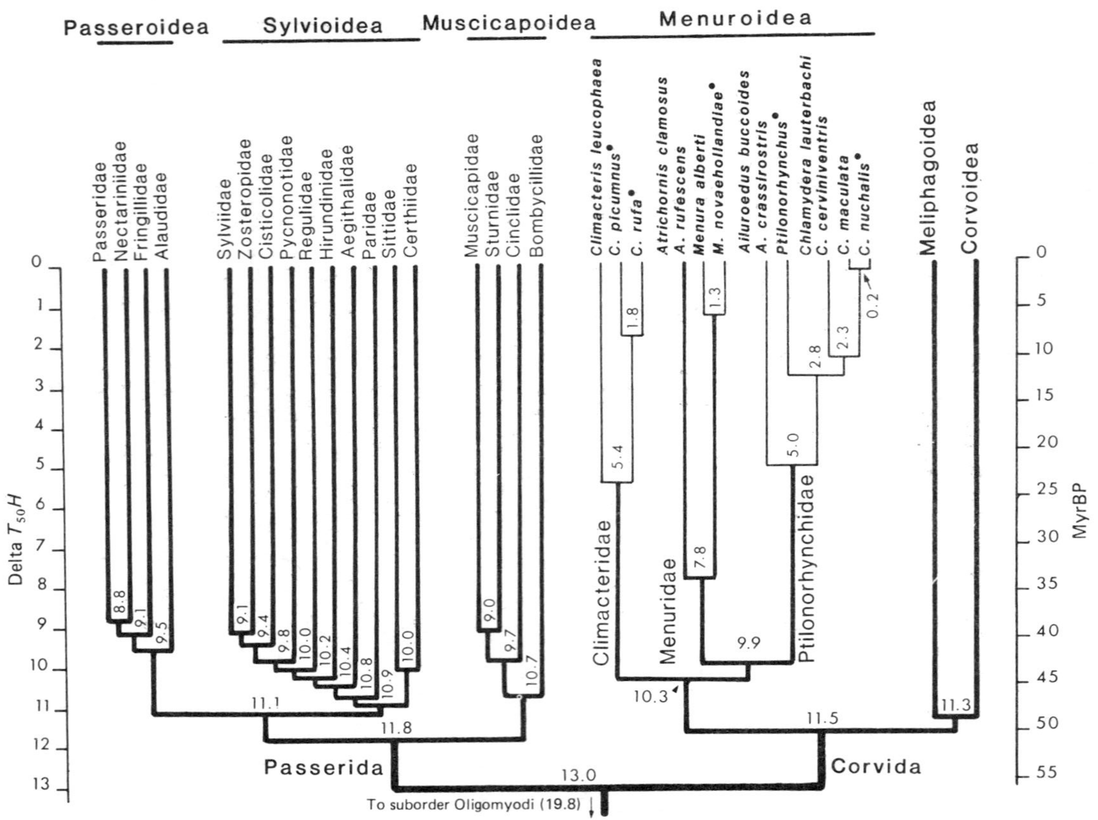

Fig. 4.4 Relationships among the superfamilies of the parvorders Corvida and Passerida. Figures 4.4 to 4.7 present the phylogeny of the Corvida of Australia and New Guinea. Oligomyodi = Tyranni (cf. p. 107).

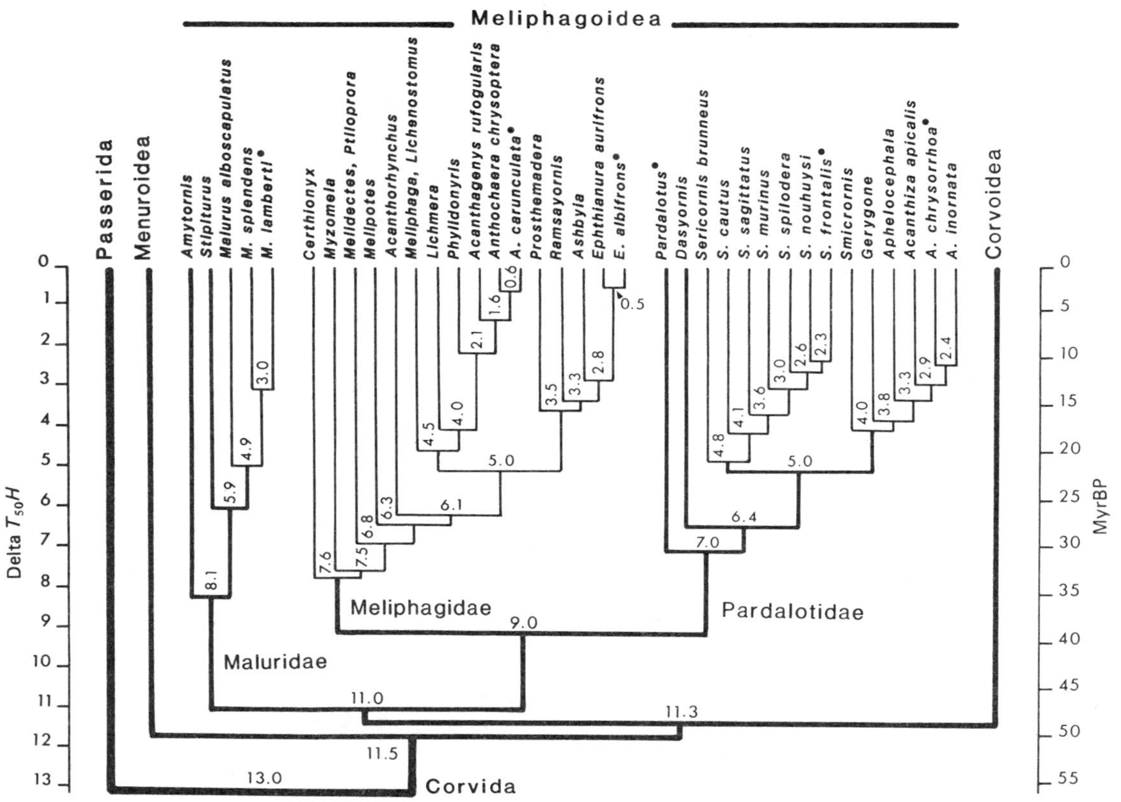

Fig. 4.5 Phylogeny of the superfamily Meliphagoidea.

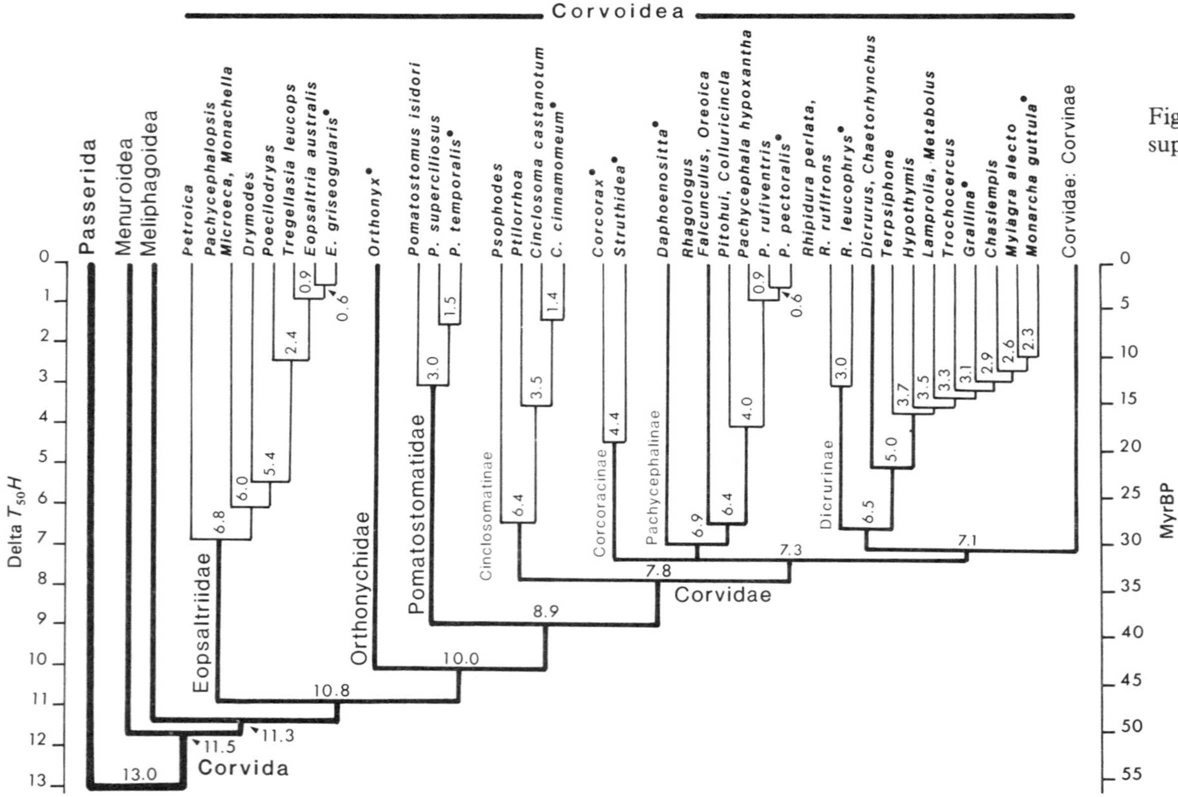

Fig. 4.6 Phylogeny of the superfamily Corvoidea.

pattern of dentition, which is why they were never confused with the placental mammals of other continents.

As Australia drifted closer to south-eastern Asia during the Tertiary, some Australian groups were able to cross the narrowing water gap and reach Asia. Among these was the ancestor of the crows, ravens, jays and magpies, the corvine birds that are so familiar to persons living in the Northern Hemisphere (Corvini, Fig. 4.7). The shrikes (Laniidae), Old World orioles and cuckoo-shrikes (Oriolini, Fig. 4.7), drongos (*Dicrurus*, Fig. 4.6), leafbirds (*Aegithina, Chloropsis*), fairy-bluebirds (*Irena*), and the New World vireos (Vireonidae) are other examples of groups with Australian ancestors.

The Bornean Bristlehead (*Pityriasis*), long a puzzle to avian taxonomists, is related to the Australian magpies, currawongs, and woodswallows of the tribe Artamini (Fig. 4.7) (Ahlquist, Sheldon & Sibley, 1984). The genus *Peltops*, of New Guinea, was often thought to be related to the monarchs, but the DNA comparisons revealed that it, too, is a member of the Artamini.

The nectar feeders (Fig. 4.5)

The nectar-feeding passerines provide an instructive example of how convergent evolution can produce similar morphologies from different ancestral lineages. Nectar feeders tend to evolve slender bills and tongues with a brush-like tip; sometimes, the tongue is extensible and it may form a tubular structure.

There are nectarivorous species in several groups, most members of which feed on fruit, seeds or insects. All or most of the species of five groups are specialized to feed on nectar. These are the honeyeaters (Meliphagidae) of Australasia, the sunbirds (Nectariniinae) of Africa and south-east Asia, the white-eyes (Zosteropidae) of Africa, south-east Asia and Australasia, the Neotropical honeycreepers (Thraupini: 'Coerebidae') of the New World tropics, and the Hawaiian honeycreepers (Carduelinae: Drepanidini) of the Hawaiian islands.

In most traditional classifications, the honeyeaters, sunbirds and white-eyes have been placed adjacent to one another, although seldom have two of them been placed in the same family. Except for their beaks and tongues, these groups are distinctive. The Neotropical honeycreepers have been proposed as the ancestors of the Hawaiian honeycreepers, although the cardueline finches have also been recognized as the probable ancestors from morphological evidence.

The DNA comparisons reveal that the Meliphagidae are one of the old endemic groups that evolved in Australia; they are members of the parvorder Corvida, thus most closely related to the other Australian endemics such as the thornbills, fairy-wrens, birds of paradise, bowerbirds, etc. (Figs 4.4 to 4.7).

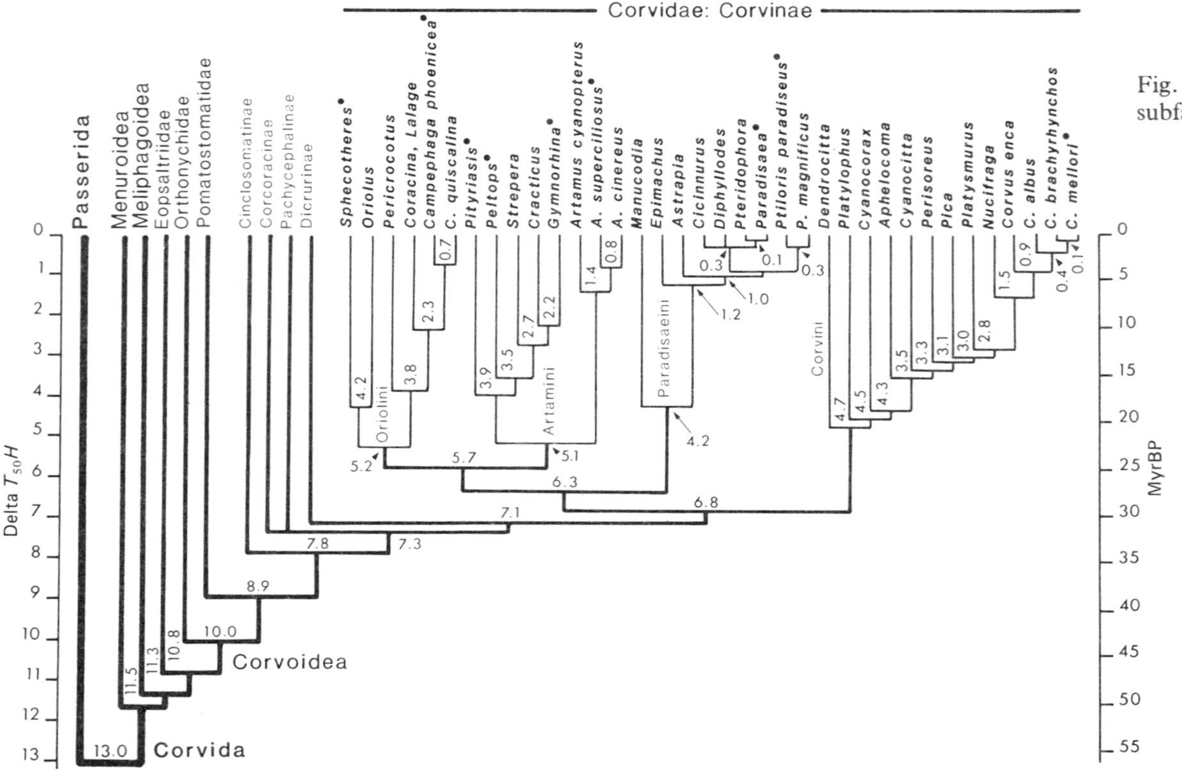

Fig. 4.7 Phylogeny of the subfamily Corvinae.

The meliphagids are more closely related to the crows and jays than to the other nectar feeders listed above.

The sunbirds are members of the Passerida, and their closest relatives include the weaverbirds (Ploceinae) of Africa and south-east Asia, and other members of the superfamily Passeroidea. The white-eyes (Zosteropidae) are also members of the Passerida, but their closest relatives are the other members of the superfamily Sylvioidea, including the sylviine warblers and babblers (Sylviinae), and the bulbuls (Pycnonotidae) (Fig. 4.4).

The Neotropical honeycreepers were usually placed in the 'Coerebidae', or thought to be allied, in part, to the tanagers, and, in part, to the New World wood warblers. The DNA data show that all of the 'coerebids' are tanagers (Fringillidae: Emberizinae: Thraupini), and that the several genera usually included in the 'Coerebidae' are not one another's closest relatives. Instead, it seems that the different genera have evolved from different ancestral tanager lineages.

The 22 species of Hawaiian honeycreepers include members with variously shaped bills, from conical seed-cracking types, and slender, pointed structures, to long, decurved bills. This variation caused early taxonomists to distribute these species among several passerine families, including the finches and the Australasian honeyeaters (Meliphagidae). Later, morphological similarities in other structures made it obvious that the Hawaiian honeycreepers are closely related to one another, but the identity of the ancestral group remained uncertain. The tanagers, cardueline finches, New World orioles (Icterini), wood warblers (Parulini) and 'coerebids' were proposed by various authors. Studies of limb musculature supported the cardueline finches as the ancestor (Raikow, 1977).

The DNA comparisons confirmed the cardueline finches as the ancestral group that provided the hardy overseas immigrant ancestor, probably from north-east Asia. The DNA data provided an approximate date for the arrival of the ancestor in Hawaii as 15–20 MyrBP (Sibley & Ahlquist, 1982).

The starlings and mockingbirds (Fig. 4.8)

One of the many surprises that have emerged from the DNA comparisons concerns the close relationship between the Old World starlings and the New World mockingbirds and thrashers. The starlings have been thought to be related to the crows, or to the weaverbirds, or to the New World troupials (Icterini). The mockingbirds and thrashers have usually been placed with, or near, the thrushes and/or the wrens.

Extensive DNA comparisons revealed that the starlings (Sturnidae: Sturnini) and mockingbirds (Sturnidae: Mimini) diverged about 25 MyrBP, and

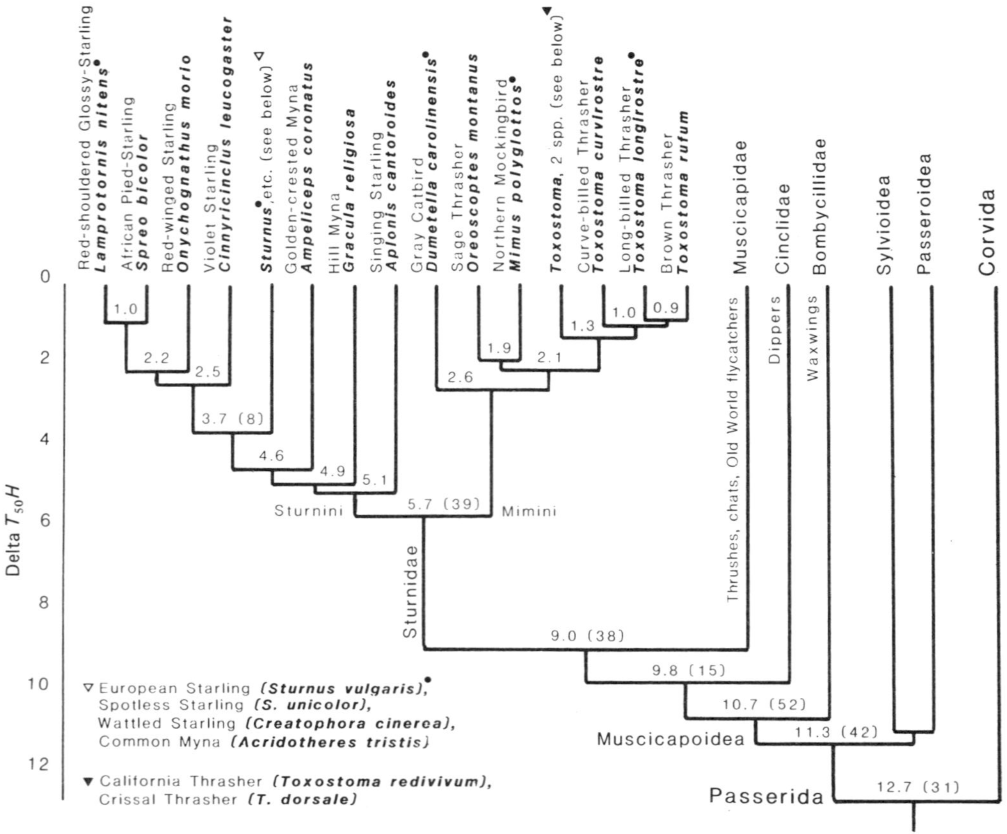

Fig. 4.8 Phylogeny of the starlings, mockingbirds and their relatives. The cluster indicated by *Sturnus*● includes four species that are approximately equidistant from the tracer species, *Lamprotornis nitens*. The relationships among the four species of starlings listed below has not been determined.

are each other's closest living relatives. They are next most closely related to the thrushes (Muscicapidae: Turdinae) and the Old World muscicapine fly-catchers and erithacine chats (Muscicapidae: Muscicapinae). If the starlings were members of the Corvida, as traditionally believed, the divergence between them and the mockingbirds would have occurred about 55–60 MyrBP (Fig. 4.4) (Sibley & Ahlquist, 1984).

A search for evidence supporting the starling–mockingbird alliance turned up a morphological study (Beecher, 1953) and a serological study (Stallcup, 1961) that also indicated a close relationship between the two groups. Comparisons of the structures of the syrinx (vocal apparatus) in the pertinent groups found that the starlings, mockingbirds and thrushes share similar syringes that differ from those of the crows and other corvines.

The starling–mockingbird relationship may be explained by the history of climatic changes in the Northern Hemisphere during the past 65 Myr. During the early Tertiary (65–30 MyrBP), a temperate climate prevailed over the Arctic, as indicated by fossils of broad-leaved trees in arctic Canada and northern Greenland (Wolfe, 1980). The common ancestor of the starlings and mockingbirds probably occurred across the entire Holarctic Region. Beginning about 30 MyrBP, the climate became colder and, by about 25 MyrBP, the ancestral population had been forced to move so far southward that it had split into Old World and New World populations, isolated by the Atlantic Ocean.

Morphology and molecules

A more general example of the congruence between morphology and DNA comparisons concerns a study (in prep.) of the structure of the head of the humerus in the Passeri. On the underside of the humerus in birds there are depressions (fossae) for the attachment of a muscle of the wing. In the Passeri there may be either one or two such fossae, depending on whether the associated muscle is or is not divided. This feature is well-known and has been considered to be unreliable as a taxonomic character because the distribution of the one- and two-fossae states is not correlated with the traditional opinions about passerine relationships (e.g. Bock, 1962). However, the single and double states are almost perfectly congruent with the phylogeny of the Passeri derived from the DNA comparisons. Nearly 99 per cent of the oscines with one fossa are members of the Corvida, and those with two fossae are members of the Passerida. The starlings, mockingbirds and thrushes have two fossae, but the crows and other Corvida have one fossa. The suboscines (Tyranni) and the non-passerines have one fossa, suggesting that this character state is primitive among birds.

Discussion

These examples illustrate several facets of the question of the relationship between morphology and molecules. Morphology has provided what appear to be the correct answers in several cases, e.g. the Hawaiian honeycreepers, and the DNA evidence always agrees with these answers. In the starling–mockingbird case, morphology had provided what is clearly a wrong answer, but when the DNA revealed the surprising truth, it was then found to be supported by both morphological and serological evidence. We have found many more examples of these kinds of situations.

We do not see a conflict between molecules and morphology; we believe that they complement one another. Comparative morphology discovered the answers to many phylogenetic questions; answers now supported by DNA comparisons. However, morphological comparisons are prone to misinterpretation and the pitfall of convergence. Special procedures will not make these problems disappear, no matter how fervent the wish of the morpho-taxonomist, or how logical the procedure employed. Morphology tracks the environment and results in adaptive modifications in structure that increase the probability of survival. It can be no other way. The result is that morphological structures are mixtures of history and function (homology and analogy), and the two are inseparably intertwined in every anatomical character. The human eye/brain mechanism is superb at recognizing *patterns*, such as human faces, but it did not evolve for the purpose of extracting the historical component from bones, muscles, behaviour, etc., in a pure form, leaving the functional, convergent, component aside. If we are ever to reconstruct phylogeny, it must be done with methods that do not rely on the human eye as the instrument of comparison. It is self-deluding to assume that what we see is all we need to know to reconstruct phylogenies. Such a procedure is subjective and qualitative, and results only in an *opinion* – and this is a definition of Art, not of Science. For a method to qualify as scientific, it must be objective and quantitative.

The proper relationship between molecules and morphology is not 'versus', but as separate methods that provide different kinds of information about evolution. The molecules can reconstruct the phylogeny with a high degree of accuracy. Given the phylogeny, the morphologist will be able to interpret structure, and to separate similarities due to common ancestry from those resulting from convergence. Our limited ability to reconstruct phylogeny from structure has prevented us from producing classifications that endure, and is the reason that systematics has been regarded as poor science by our colleagues in other disciplines. We can change this if we will cease the debate about 'molecules *vs* morphology' and begin to co-operate on the basis of molecules *and* morphology.

This generation of systematists is fortunate: we are the first to be provided with the opportunity to use objective, quantitative methods for the reconstruction of phylogeny, and to develop classifications based on phylogeny. Those who employ such methods will experience the pleasure and satisfaction of correcting past errors, making unexpected discoveries, and setting the stage for the future of evolutionary biology.

REFERENCES

Ahlquist, J. E., Sheldon, F. H. & Sibley, C. G. (1984). The relationships of the Bornean Bristlehead (*Pityriasis gymnocephala*) and the Black-collared Thrush (*Chlamydochaera jefferyi*). *Journal für Ornithologie*, **125,** 129–40.

Ames, P. L. (1971). The morphology of the syrinx in passerine birds. *Bulletin of the Peabody Museum of Natural History*, **37,** 1–194.

Beecher, W. J. (1953). A phylogeny of the oscines. *The Auk*, **70,** 270–333.

Bock, W. J. (1962). The pneumatic fossa of the humerus in the Passeres. *The Auk*, **79,** 425–43.

Cottam, P. (1957). The pelecaniform characters of the shoe-bill stork, *Balaeniceps rex. Bulletin of the British Museum (Natural History), Zoology*, **5,** 51–72.

Heimerdinger, M. A. & Ames, P. L. (1967). Variation in the sternal notches of suboscine passerine birds. *Postilla*, **105,** 1–44.

Olson, S. & Feduccia, A. (1980). Relationships and evolution of flamingos (Aves: Phoenicopteridae). *Smithsonian Contributions to Zoology*, **316,** 1–73.

Raikow, R. J. (1977). The origin and evolution of the Hawaiian honeycreepers (Drepanididae). *Living Bird*, **15,** 95–117.

Sibley, C. G. & Ahlquist, J. E. (1981). The phylogeny and relationships of the ratite birds as indicated by DNA–DNA hybridization. In *Evolution Today, Proceedings Second International Congress of Systematic and Evolutionary Biology*, ed. G. G. E. Scudder & J. L. Reveal, pp. 301–35. Pittsburgh, Pennsylvania: Hunt Institute for Botanical Documentation, Carnegie-Mellon University.

Sibley, C. G. & Ahlquist, J. E. (1982). The relationships of the Hawaiian honeycreepers (Drepaninini) as indicated by DNA–DNA hybridization. *The Auk*, **99,** 130–40.

Sibley, C. G. & Ahlquist, J. E. (1983). Phylogeny and classification of birds based on the data of DNA–DNA hybridization. In *Current Ornithology*, vol. 1, ed. R. F. Johnston, pp. 245–92. New York: Plenum Press.

Sibley, C. G. & Ahlquist, J. E. (1984). The relationships of the starlings (Sturnidae: Sturnini) and the mockingbirds (Sturnidae: Mimini). *The Auk*, **101,** 230–43.

Sibley, C. G. & Ahlquist, J. E. (1985*a*). The phylogeny and classification of the Australo–Papuan passerine birds. *The Emu*, **85,** 1–14.

Sibley, C. G. & Ahlquist, J. E. (1985*b*). The relationships of some groups of African birds, based on comparisons of the genetic material, DNA. In *Proceedings of the International Symposium on African Vertebrates, Bonner Zoologische Beitrage*, ed. K.-L. Schuchmann, pp. 115–61. Bonn: Museum Alexander Koenig.

Sibley, C. G. & Ahlquist, J. E. (1985*c*). Phylogeny and classification of the New World suboscine passerine birds (Passeriformes: Oligomyodi: Tyrannides). In *Neotropical Ornithology*, Ornithological Monographs No. 36, ed. P. A. Buckley, M. S. Foster, E. S. Morton, R. S. Ridgely, & F. G. Buckley, pp. 396–428. Washington, D.C.: American Ornithologists' Union.

Stallcup, W. B. (1961). Relationships of some families of the suborder Passeres (songbirds) as indicated by comparisons of tissue proteins. *Journal of the Graduate Research Center, Southern Methodist University*, **29**, 43–65.

Wolfe, J. A. (1980). Tertiary climates and floristic relationships at high latitudes in the northern hemisphere. *Palaeogeography, Palaeoclimatology, & Palaeoecology*, **30**, 313–23.

APPENDIX

Details of the DNA–DNA hybridization technique

Long-chain DNAs are extracted from the nuclei of cells, purified to remove proteins and RNA, and sheared by high-frequency sound waves (sonication) into fragments with an average length of 500 bases. The single-stranded fragments of the species to be radio-labelled and used as 'tracers' are allowed to reassociate for about 2 hours at 50 °C in 0.48 M sodium phosphate buffer. This permits most of the rapidly reassociating repeated sequences to form duplexes, while the slowly reassociating single-copy sequences remain single-stranded. The latter are recovered by chromatography on a hydroxyapatite (HAP) column. (In 0.12 M sodium phosphate buffer, HAP, a form of calcium phosphate, binds double-stranded but not single-stranded DNA.) This produces a single-copy preparation consisting of one copy per genome of each original single-copy sequence, plus *at least* one copy per genome of each repeated sequence. Thus, the so-called 'single-copy' DNA preparation actually contains representatives of all of the different DNA sequences in the nucleus of each cell. The single-copy DNA is then labelled with radioiodine (^{125}I). DNA–DNA hybrids are formed from one part of the radioactive tracer DNA and 1000 parts of each of the 'driver' DNAs of the species that are to be compared with the tracer species. The 1 : 1000 ratio between tracer and driver increases the probability that tracer–driver hybrids will be formed, and reduces the probability that tracer–tracer hybrids will be formed. One of the combinations in each set consists of tracer and driver DNAs from the species used as the tracer. This is the homoduplex hybrid which is used as the basis for comparisons with the various tracer–driver hybrids. The hybrid combinations are boiled for 5 minutes to dissociate duplexes into simplexes, then incubated for 120 hours at 60 °C in 0.48 M phosphate buffer to permit the single strands to form duplexes. After incubation, the hybrids are placed on HAP columns in 0.12 M phosphate buffer in a temperature-controlled waterbath at 55 °C. The temperature is then raised in 2.5 °C increments from 55 to 95 °C. At each of the 17 temperatures, the single-stranded fragments produced by the melting of hybrid duplexes are eluted in 20 ml of 0.12 M phosphate buffer. The

radioactivity in each sample is counted and the values are used to construct the melting curves and to calculate the median melting temperature differences between the homoduplex hybrid and each of the heteroduplex hybrids.
See Sibley & Ahlquist (1983) for additional details.

5 Tetrapod relationships: the molecular evidence

M. J. Bishop & A. E. Friday

Introduction

Patterson (1980) has reviewed the history of the concept of the Tetrapoda, and demonstrated that the discovery and description of the lungfishes (Dipnoi) provoked discussion of issues which cannot be said to have been resolved satisfactorily. He shows how early opinion was in favour of the interpretation that lungfishes were either the ancestors of, or the nearest relatives to, the tetrapods. For some (Watson, 1926; Westoll, 1943), this interpretation gave way to the view that the tetrapods (as a monophyletic group) were the descendants of the osteolepid rhipidistians. To others (Holmgren, 1933; Jarvik, 1942), the tetrapods did not appear to be a monophyletic assemblage. Jarvik (1980) would derive the living urodeles from porolepiform rhipidistians and the other tetrapod groups from osteolepiforms.

Patterson demonstrates that the question of tetrapod ancestry depends for its solution upon our ability at least to be able to characterize the tetrapods as an assemblage. With others (Rosen *et al.*, 1981), Patterson concludes from a cladistic analysis that the hypothesis which accepts the lungfishes as the closest living relatives of the tetrapods is the best supported by synapomorphies. He regards the rhipidistians as a paraphyletic group and points out that we lack any credible phylogeny of the tetrapod groups. He also disagrees with some of the characters used by Gaffney (1979*a, b*) to support tetrapod monophyly.

Holmes (1985), in a review of the paper by Rosen *et al.* (1981), takes issue with the conclusion that lungfishes and tetrapods are sister-groups. He argues that a more detailed consideration of intragroup variation, and a scheme of weighting proposed synapomorphies, substantially weakens the case for the lungfish–tetrapod grouping. In particular, he suggests that many of the synapomorphies used by Rosen *et al.* are of low weight because of a high probability of independent origin. However, Holmes agrees that there are indeed weaknesses in the orthodox view of rhipidistian–tetrapod relationships, and that a thorough re-examination of tetrapod origins is warranted.

Recent opinions on tetrapod interrelationships

In recent years there have been several attempts to re-examine tetrapod interrelationships. Løvtrup (1977) presented a phylogenetic classification of the vertebrates based largely on a range of developmental, physiological and biochemical characters, some of which are sufficiently unfamiliar that they have yet to be evaluated by other authors. Løvtrup (1977, p. 186, fig. 4.18) displays a phylogenetic classification in which the tetrapods constitute a monophyletic assemblage, but within the tetrapods the living amphibian groups are not considered to comprise a monophyletic assemblage. He also suggests that the Chelonia be considered as sister-group to the Crocodylia and Aves, with the Rhynchocephalia (the living *Sphenodon*) as sister-group to these two. The Squamata are then shown as sister-group to the Crocodylia plus Aves plus Chelonia plus Rhynchocephalia, and the Mammalia are represented as sister-group to all these (Fig. 5.1*a*). More recently, however, Løvtrup (1986) has inclined to the view that birds and mammals should be regarded as sister-groups, a view discussed further below.

In a review of the interrelationships of jawed vertebrates, Wiley (1979) presented a new classification of the Vertebrata in which he accepted that the Amphibia are monophyletic and are sister-group to the Amniota in a monophyletic assemblage of tetrapods. For the amniotes, the structure of his classification mirrors the generally accepted view that birds are most closely related to crocodilians among living groups, and that the relationships between mammals, lepidosaurs (snakes, lizards, *Sphenodon*) plus archosaurs (birds, crocodiles), and chelonians (turtles) are uncertain.

Perhaps the most radical (and controversial) recent reanalysis of tetrapod interrelationships is that of Gardiner (1982). Gardiner fully accepts tetrapod monophyly (his p. 208: 'That the tetrapods are a natural assemblage cannot be doubted. The pentadactyl limb with its carpus, tarsus and dactyly convinces us of this fact.') He argues that the living Amphibia do constitute a monophyletic group and that this is the sister-group to the Amniota. Within the tetrapods, he concludes that the birds are the sister-group of mammals; crocodiles are the sister-group of those two; chelonians the sister-group of those three; and squamates (snakes, lizards) plus *Sphenodon* the sister-group of those four (Fig. 5.1*b*).

The most surprising feature of Gardiner's analysis has been his discounting of the evidence for a synapsid–mammalian transition. This transition has been generally regarded as well documented and as a paradigm for the origin of a major type of tetrapod body-plan. To quote Gardiner:

The detailed correspondence between mammals and birds far outweighs the evidence of a synapsid–mammalian transition. The Synapsida, as befits a paraphyletic group, lacks synapomorphies, whereas mammals and birds share derived features of endothermy, thermoregulation, form of the heart

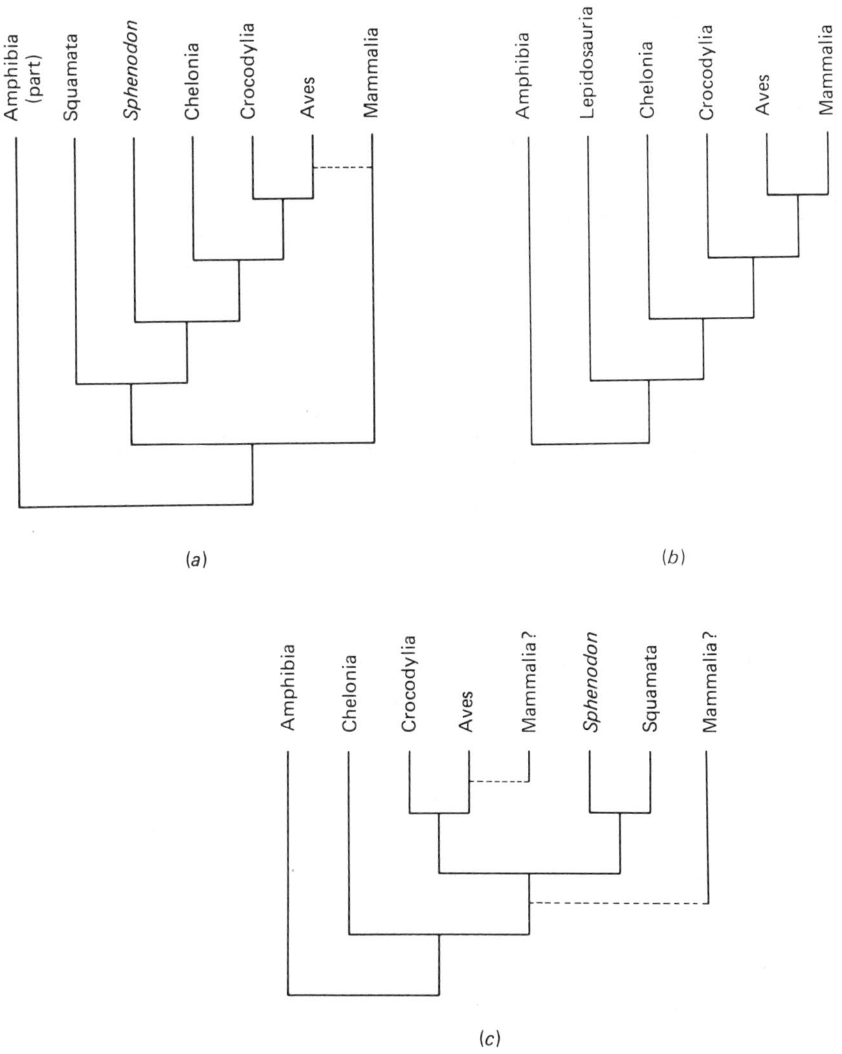

Fig. 5.1 Trees depicting relationships of major tetrapod groups as expressed by authors quoted in the text. (*a*) simplified after Løvtrup (1977) – the hatched line represents Løvtrup's (1986) assessment of the position of the mammals; (*b*) redrawn after Gardiner (1982); (*c*) composite tree from diagrams and text of Benton (1985) – the hatched lines reflect two possible positions for mammals.

> and separation of the lung from body circulation, meninges, cerebellum, turbinals, maxillary process, adventitious cartilage, atlas-axis, vascularized islets in the pancreas, pinealocytes, macula densa, loop of Henlé and LDHX.
>
> Moreover, when the Synapsida were broken down into some of the constitutent monophyletic groups none of these . . . was found to possess a single derived feature in common with the Mammalia alone.

Gardiner's character-state tree of the tetrapods (Fig. 5.1*b*) thus differs from Løvtrup's (Fig. 5.1*a*) conclusions principally in the position of the mammals. Løvtrup, in 1977, reflected the widely held view that the living reptile groups plus the birds constitute a monophyletic assemblage the ancestry of which diverged from the lineage leading to mammals. It is, incidentally, quite clear from his text that Løvtrup considered his classifications in an evolutionary context (for example, his p. 184: 'The conclusion that Mammalia represent a separate evolutionary line therefore seems unavoidable . . .'). As noted above, Løvtrup has recently changed his conclusions to support a grouping of mammals with birds. Gardiner, indeed, unites birds and mammals as sister-groups in the taxon Haemothermia (Owen, 1866). It should be emphasized that Richard Owen, who opposed Darwin's theory of evolution, was under no obligation to regard his classification as a reflection of the results of an evolutionary process.

Gardiner's account should be consulted for a historical introduction to the problems of tetrapod classification, particularly the provenance of some of the characters used by Goodrich, Cope, Broom, Haeckel, Huxley, Lancaster, Watson and others to support the classic case for a twofold division of all amniotes into the major types, Sauropsida (Chelonia, Squamata, Rhynchocephalia, Crocodylia and Aves) and Theropsida (Mammalia), in opposition to Owen's Haemothermia.

It is clear from a reading of Holmes (1975) (quoted in support of the Haemothermia by Gardiner (1982)) that the resemblances of the bird heart to the mammal heart are extraordinarily detailed. If these features are not synapomorphies of the Haemothermia, they are very impressive examples of convergence. Either way, these characters surely deserve further study. Holmes's work also serves to emphasize the variety of anatomical and physiological adaptations now beginning to be revealed in the heart structure and circulation pattern of the groups of living reptiles.

It has also been pointed out, again by Løvtrup (1977, p. 185), that:

> . . . a number of similarities exist between Crocodylia and Mammalia. These are the subdivision of the vertebral column into five distinct regions, a secondary palate and thecodont teeth. In addition, we have, of course, the four-chambered heart and a

complete diaphragm, the presence of a cochlea in the ear and of a true cerebral cortex. All these features must [Fig. 5.1*a*], according to the proposed classification, be accepted as instances of convergence.

If birds and crocodiles are indeed descendants of a common archosaur stock, then perhaps some of the resemblances between both groups and the mammals may be more readily understood. A case has been made by Reig (1967) that the archosaur groups, including the living crocodiles and birds, actually arose from the pelycosaurs, usually accepted as the ancestors of other mammal-like reptiles (and hence, eventually, of the mammals). The lepidosaur groups, including the living squamates (lizards and snakes) and rhynchocephalians, are derived by Reig from the captorhinomorph cotylosaurs (specifically 'through the Millerettiformes'). Occasional doubt has been expressed, in any case, over the relevance of pelycosaurs to therapsid ancestry (for example, by Gardiner (1982)), but Reig's hypothesis has been poorly received for other reasons. These reasons are brought out best in Benton's (1985) recent review of the classification and phylogeny of the diapsid reptiles.

Benton directs much attention to the attacks by both Løvtrup (1977) and Gardiner (1982) (and earlier by Romer (e.g. 1971)) on the monophyly of the Diapsida. He presents a historical survey and then examines the evidence character by character. Benton's own conclusion, and that of several other recent authors quoted by him (for example, Gaffney (1980) and Reisz (1981)) is that the diapsid reptiles do constitute a monophyletic assemblage; that, among living forms, crocodiles (plus birds) are the sister-group of lepidosaurs rather than of chelonians; that chelonians are then the sister-group to these two (Fig. 5.1*c*); and that (his p. 107): 'The out-group for this section is living amphibians (non-amniote tetrapods), on the assumption that the Tetrapoda are monophyletic.'

Benton does not consider the vexed question of the position of the mammals, although he includes birds as diapsids where relevant to his character analyses. He concludes that all diapsids later than the Upper Carboniferous *Petrolacosaurus* (the oldest recorded diapsid) fall into two large groups: the Division Archosauromorpha (including the living crocodiles and birds) and the Division Lepidosauromorpha (including the living squamates and *Sphenodon*). It should be noted that Benton has found no close relationship between the Triassic rhynchosaurs and *Sphenodon*: the rhynchosaurs are on the archosauromorph side of the diapsid divide.

Benton has also considered molecular data, and his account may be consulted for a summary of the conflicting published analyses, some of which will be referred to below. Following Romero-Herrera *et al.* (1978), he emphasizes the lack of convincing differences between parsimony solutions for competing phylogenetic hypotheses, and the possibility of convergent evolutionary change at the molecular level.

Molecular evidence

Molecular sequence data are lacking or very sparse for a number of tetrapod groups. Mammals are quite well served (see McKenna, this volume, pp. 55–75), as one might expect from our 'theriocentric' point of view. Often, however, a particular reptilian species will be compared with mammals, for example in the study of the caiman and mouse immunoglobulin V_H multigene family (Litman *et al.*, 1983). Without further comparisons, it is not possible to evaluate the significance of some striking similarities between the animals: perhaps all tetrapods or all amniotes will show a similar organization of the genome for the features under consideration.

In some cases physiological investigations can be viewed in the light of known features of molecular structure. For example, the extensive information about the three-dimensional configuration of haemoglobins has been used to examine the significance of the observation that bicarbonate ions lower the oxygen affinity of the haemoglobins of crocodilians (Perutz *et al.*, 1981). For myoglobin, Watts, Angelides & Brown (1983) determined the amino acid sequence of that protein from the Pacific green sea turtle (*Chelonia mydas caranigra*) and compared the molecule with the myoglobins of another chelonian, a lizard, an alligator and mammal, bird, two shark and two teleost fish sequences. Watts *et al.* (1983) compared all known vertebrate myoglobins and their analysis supports the common origin of the tetrapods on the grounds that the tetrapod sequences share a number of features believed to be derived in comparison with those of non-tetrapods. They note, for example, that residues 133 to 139 (helical notation H10–H16) are invariant in known tetrapod myoglobins, yet this region is variable in the cartilaginous and bony fish myoglobins.

This sort of evidence suggests that the oxygen-binding proteins of tetrapods have come under new constraints in the terrestrial environment, so that new parts of the molecules have been severely conserved by natural selection. Region 133 to 139 of myoglobin was the subject of comment by Romero-Herrera *et al.* (1978) who suggested that those residues on the inside of the molecule in this region may be involved in contact with the haem. It was also suggested that the nucleotides coding amino acids 133 to 139 in the mRNA might be involved in a hair-pin loop with the nucleotides coding amino acids 152 to 146. The recently published nucleotide sequence of the myoglobin gene of the seal (Blanchetot *et al.*, 1983) enables this hypothesis to be further examined. There appear to be 9 base pairings (in a proposed stem 17 bases long) if these two regions are aligned, somewhat less than the 15 hypothesized by Romero-Herrera *et al.* (1978). We therefore suggest that the existence of a hair-pin loop in the myoglobin mRNA is unlikely.

There may also be physiological investigations of animals which reveal characteristics of interest, but without corresponding sequence information as yet for those animals for the molecule involved. An example is the study of the co-operative haemoglobin–oxygen binding in *Sphenodon* (Wells, Tetens & Brittain, 1983). *Sphenodon* was found to be unique among terrestrial vertebrates so far studied, but no sequence is yet available for any *Sphenodon* haemoglobin chain to enable the physiological properties to be tied to particular features of the protein structure.

It is not always easy to suggest a functional role for a particular part of the myoglobin molecule. Dene *et al.* (1980) reported the sequence of American alligator (*Alligator mississippiensis*) myoglobin, and in a parsimony analysis of mammal, bird, chondrichthyan and teleost sequences found that the mammals and birds joined as sister-groups, with the alligator then joining mammals plus birds. Forcing a more conventionally acceptable pattern with birds and alligator as sister-groups was less parsimonious. The difference was relatively small (8 nucleotide replacements or NRs) compared to the amount of reconstructed change over the whole tree (630 NRs). Five amino acids shared by the bird and mammal lineages had to be reconstructed as parallel substitutions when these lineages were separated in the conventional pattern of relationships (Fig. 5.2). These could not be assigned a functional role when viewed in the light of what is known about the functioning of the myoglobin molecule, although residue 40 (leucine) is in a section of the myoglobin molecule which contains a number of residues which change only in a very conservative fashion, if at all, in known myoglobins (Romero-Herrera *et al.*, 1978), and it is not impossible that residue 40 may prove to be of functional significance.

When Dene *et al.* (1980) considered the implications of the alligator sequence, their sample of bird myoglobins comprised only the chicken and penguin. There has been a revision of the chicken myoglobin sequence originally determined by Deconinck *et al.* (1975). Prass, Berkley & Romero-Herrera (1983) found no less than 11 differences from the originally determined sequence, and we use their sequence for chicken cardiac myoglobin. One of the residues involved in the revision (glutamine for histidine at position 81) was one of the five residues noted previously by Dene *et al.* as parallel changes between birds and mammals if these groups were separated on the tree. There are now, therefore, only four such residues as far as the chicken is concerned. The penguin myoglobin sequence determined by Peiffer (1973) has been omitted from our comparisons because there is some doubt over some of the positions.

Maeda & Fitch (1981*a, b*) reported the amino acid sequences for the myoglobins of map turtle (*Graptemys geographica*) and lace monitor lizard

(*Varanus varius*), and used a maximum parsimony approach to review the topologies of trees determined on the basis of myoglobin, alpha haemoglobin and cytochrome *c* for a variety of tetrapods. Their analysis was inconclusive regarding the interrelationships of tetrapod taxa, but their most parsimonious tree for myoglobin links birds and mammals as sister-groups, in agreement with the finding of Dene *et al.* (1980). However, Maeda & Fitch (1981*b*) emphasize the rapid succession of divergences leading to the lineages of the major amniote groups: they interpret the evidence from the fossil record as indicating that the chelonian–diapsid–synapsid divergences occurred over a period of only some 10 Myr, 300 MyrBP. If true, this is not an auspicious set of conditions for the preservation of unequivocal information in the molecular sequences which survive and are available for sampling today.

Maximum likelihood estimation of evolutionary trees

There are relatively few sets of sequence data that contain information from representatives of sufficient of the various tetrapod groups to make them of value for the present study. Nucleic acid sequence data have been

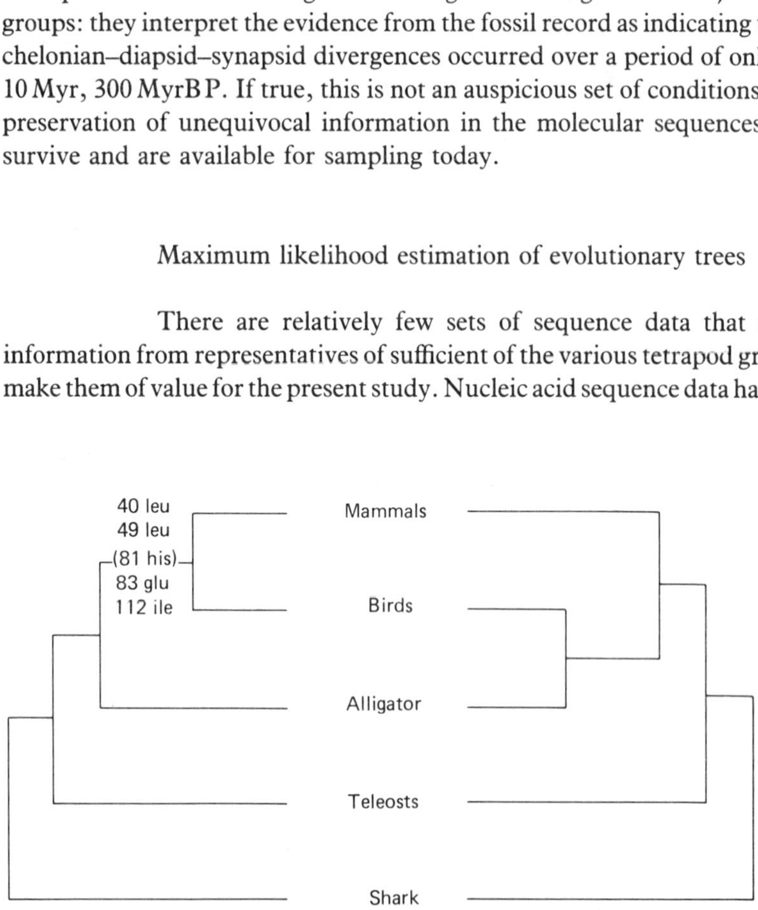

Fig. 5.2 Dene *et al.* (1980) constructed a maximum parsimony tree for the myoglobin sequences of 49 mammals, 2 birds, an alligator, 2 bony fishes and a shark. That tree is given on the left of the figure, with five amino acids shared by the mammals and birds shown on their common stem. In the more conventional grouping, given on the right, it would be necessary to reconstruct the five amino acids shared by mammals and birds as parallel substitutions, a less parsimonius solution. Since Dene *et al.* published their analysis, corrections have been made to the chicken sequence (see text), which now has gln at position 81 instead of his.

determined for the most part without possible phylogenetic comparisons in mind.

As far as protein sequences are concerned, reptile and amphibian groups are poorly represented again in all but globin sequences (see Goodman, Miyamoto & Czelusniak, this volume, pp. 173–6). There are three crocodilian beta haemoglobin sequences but other reptile groups are not represented for this protein, although there are several sequences from anurans. We have, therefore, turned to myoglobin and alpha haemoglobin in this study.

For myoglobin there are many mammal sequences (see McKenna, this volume, pp. 58–68), two bird sequences, two for chelonians, one lizard and one alligator sequence. So far, there are no myoglobin sequences available for amphibians, although a variety of cartilaginous and bony fishes are represented. For alpha haemoglobin there are numbers of mammals and birds represented plus three crocodilian sequences, a snake sequence, three amphibian sequences and those for several bony fishes. All our previous analyses of sequence data indicate that the amphibians can be regarded as the sister-group to the Amniota, and this has enabled us further to concentrate on relationships within the Amniota.

Our phylogenetic estimation has been carried out under a probabilistic model of change at the level of DNA sequences. A protein sequence can, with uncertainty due to the redundancy of the genetic code (and hence ambiguity in translating amino acids into nucleotides), be expressed as a DNA sequence. We view a nucleic acid sequence as a family of independent, identically distributed random variables (one for each site) which have the Markov property, i.e. conditional independence of the next transition and the past transition given the present state. We assume that the changes inferred in the corresponding DNA sequences from the observed differences between protein sequences arrive according to a Poisson process. The rate of substitution in the DNA is assumed to be the same for all sites. This enables us to compute the probability of substitution over a given time interval, as follows. A given site may be in any of the states A, C, G, T. When a substitution occurs, let a base be replaced by A, C, G or T with equal probability ($\frac{1}{4}$). In terms of a Poisson process, the probability $(1 - e^{-ut})$ that substitution will occur in time t (at rate u) implies the probability e^{-ut} that it will not. For any site, the probability of finding base X at time 0 and a different base Y at time t is thus $(1 - e^{-ut})/4$, and the probability of finding the same base X after time t is $(1 + 3e^{-ut})/4$. A model of this form was proposed by Felsenstein (1981).

We (Bishop & Friday, 1985) have reviewed the development of probabilistic models of evolutionary change at the molecular level, and described in detail the approach we take here to estimating phylogenetic trees from nucleic acid and protein sequence data. The maximum likelihood method of estimation involves computing probabilities of change over all the trial tree patterns under the model of change. The *support* value for a tree is the natural logarithm of the likelihood. The method is quite expensive in terms of computing time: for example, to examine the myoglobin data for six species discussed below took some 40 minutes of central processor unit time on an IBM 3081 computer. In any case, the problem of finding that tree pattern for which the likelihood is globally a maximum is intractable, unless one examines all possible trees for a given number of species, a task that becomes computationally impossible (for around 10 or more species). Hence we take several approaches to estimating a maximum likelihood tree without examining all possible tree patterns for a given number of species.

First, we can take each pair of sequences and use these in turn to generate all the pairwise estimates of relative divergence time (see Bishop & Friday (1985) for the derivation of a divergence time estimator based on a Poisson process of change). The matrix of divergence time estimates can then be clustered using the unweighted pair-group algorithm to give the pairwise estimate of the tree. This method is a fairly coarse one but has the advantage of being computationally feasible for large numbers of sequences.

A more rigorous approach to using the information in the sequence data is to attempt joint estimation of the tree. Here the estimated divergence times for a particular tree pattern are refined in turn over many cycles of iteration. As one time is altered in the light of the others, the others in turn can be further adjusted until the value of the likelihood computed over the whole tree using those times is at a maximum.

We carry out joint estimation in two ways, either by starting with all species originating at a common time in the past (the 'big-bang' pattern of Thompson (1975)) and then expanding that pattern progressively into a dichotomous one, or by submitting hypotheses of tree patterns based on published studies of the groups using comparative morphological or other data.

All these approaches have been applied to a restricted set of myoglobin sequences, and Fig. 5.3 shows results of likelihood estimation based on the myoglobins of the six species *Homo sapiens* (man), *Didelphis marsupialis* (opossum), *Gallus gallus* (chicken), *Alligator mississippiensis* (alligator), *Varanus varius* (lace monitor lizard) and *Graptemys geographica* (map turtle). References to the sources of the sequences for man, opossum and chicken are given in Romero-Herrera *et al.* (1978); the alligator sequence is taken from Dene *et al.* (1980); the lace monitor lizard sequence is from Maeda & Fitch (1981*b*) and the map turtle sequence is from Maeda & Fitch (1981*a*).

In Fig. 5.3 the support value for each pattern is shown. In the case of many other topologies the trial times of different nodes collided during iteration. When such collision of nodes occurs, the pattern as submitted to the procedure is unstable and the support value has no meaning. In these cases that particular pattern is rejected and new patterns are submitted to the procedure, both the collapsed pattern with one of the superimposed nodes eliminated and a pattern in which the order of the superimposed nodes is reversed.

The support values may be compared with one another and each may also be compared to the support value for the 'big-bang' pattern. If no pattern is more than trivially supported over the 'big-bang', then the data do not carry sufficient information to enable a choice to be made about the order of branching. For the data on which Fig. 5.3 is based the support value for the 'big-bang' pattern is −2255.707.

For the myoglobins of the six species considered (Fig. 5.3), patterns 5.3*a* and 5.3*b* are best supported. However, although they are substantially supported over the 'big-bang' pattern, they differ very little in support from one another, and also differ very little from patterns 5.3*c* and 5.3*d*. Pattern 5.3*e* is the pattern of next highest likelihood, and is already about six support units away from the maximum likelihood solution, tree 5.3*a*. All other patterns examined (some 40 were evaluated) are 12 or more support units from the maximum.

Pattern 5.3*b*, one of the two best-supported patterns, corresponds with the pattern estimated using the pairwise procedure. The relative times estimated by pairwise (Fig. 5.3*b*) and joint (Fig. 5.3*a*) procedures also differ little. For these data, therefore, the computationally much less complicated and less time-consuming pairwise procedure has performed well. This is not always found to be the case, but is to be expected when the underlying model is appropriate to the analysis of the data. (Remember that our model assumes stochastically constant rates of change.) We also examined a much larger set of 61 myoglobin sequences by the pairwise procedure, and the pattern for the major groups of amniotes was also as shown in Fig. 5.3*b*.

Some of the patterns evaluated during estimation of maximum likelihood trees correspond to the opinions of various authors discussed above. In all cases, except two, times collided and rearrangement of the topology was indicated (none of the patterns shown in Fig. 5.1 was stable). The exceptions are patterns 5.3*a* and 5.3*b*, the two best-supported, which correspond with the alternative, equally most parsimonious patterns found by Maeda & Fitch (1981*b*). These patterns reflect the uncertainty of the placing of alligator and lizard relative to turtle. It seems, therefore, that our analysis and that of Maeda & Fitch (1981*b*), using different models of change, both favour the immediate common ancestry of the two mammals and one bird. On the other hand, both approaches emphasize the uncertainty of the interrelationships of turtle, lizard and alligator. The pairing lizard–alligator is not favoured, but neither the turtle–lizard nor the turtle–alligator pairing is highly supported over the other.

The relationships between mammals, birds and crocodilians were further explored using two sequences of alpha haemoglobins from each of these groups. The sources of the data are given in Perutz *et al.* (1981) and the results

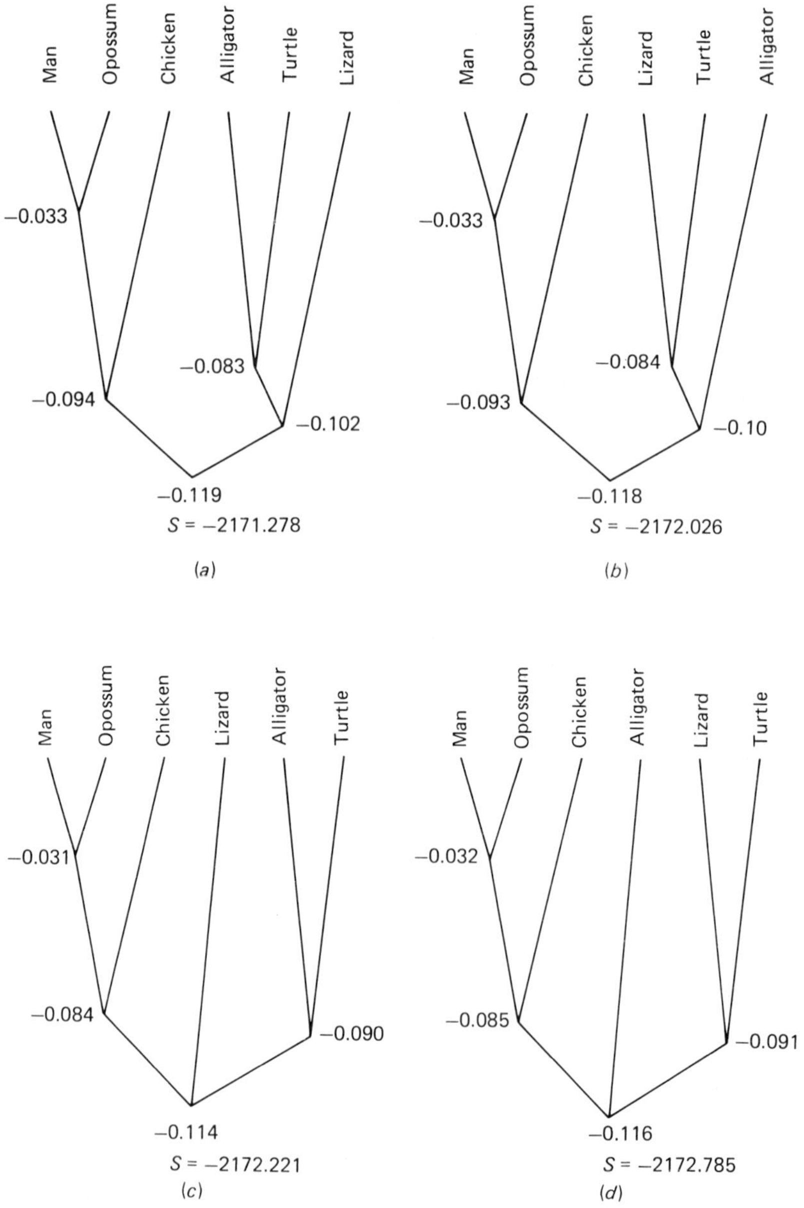
Man
Opossum
Chicken
Alligator
Turtle
Lizard
−0.033
−0.083
−0.094
−0.102
−0.119
S = −2171.278
(a)
Man
Opossum
Chicken
Lizard
Turtle
Alligator
−0.033
−0.084
−0.093
−0.10
−0.118
S = −2172.026
(b)
Man
Opossum
Chicken
Lizard
Alligator
Turtle
−0.031
−0.084
−0.090
−0.114
S = −2172.221
(c)
Man
Opossum
Chicken
Alligator
Lizard
Turtle
−0.032
−0.085
−0.091
−0.116
S = −2172.785
(d)

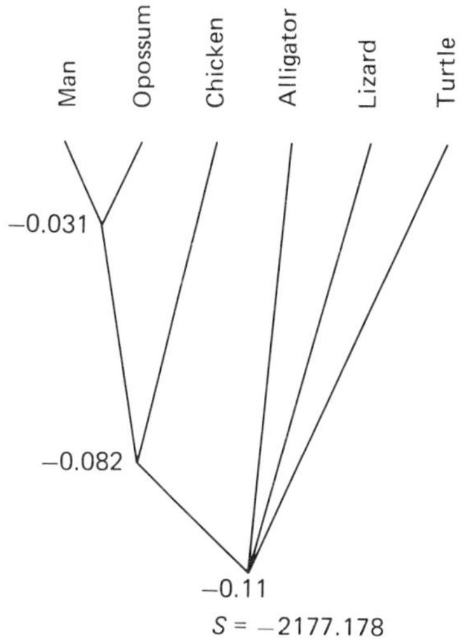

(*e*)

Fig. 5.3 Maximum likelihood estimates of trees based on myoglobin sequence data for six species of tetrapods referred to in the text. Relative estimates of time ago are given against each branch point. For each tree these may be directly converted into estimates of absolute time if one calibrated point is regarded as established. The support value, S, associated with each tree is the log likelihood for that tree, under the model of change adopted (see text). Tree *a* has the highest likelihood found, trees *b*, *c*, and *d* are of only slightly lower likelihood. Tree *e* is some 6 support units lower than the maximum likelihood solution, *a*; all other patterns examined were 12 or more units lower than tree *a*.

are shown in Fig. 5.4 The support value for the 'big-bang' pattern is −2235.114. The two patterns 5.4*a* and 5.4*b* can hardly be distinguished in terms of their support values, yet one (5.4*a*) represents the viewpoint of Gardiner (1982), with birds and mammals as sister-groups, whereas the other (5.4*b*) represents the conventional grouping of birds with crocodilians. Pattern 5.4*c* has a similar support value, reflecting the relatively inconclusive nature of this evidence from the alpha haemoglobins. The grouping of mammals plus crocodilians is not a stable one for these data. Again, the pairwise method performed well and estimated a tree identical to that shown in Fig. 5.4*b* in pattern, and very close to that solution in the times assigned to branching points.

It would be premature to conclude that the available molecular sequence data necessarily threaten the conventionally long-separate phyletic histories of mammals and birds. Both myoglobin and haemoglobin are involved in oxygen storage and transport, and it could well be that the relatively high metabolic

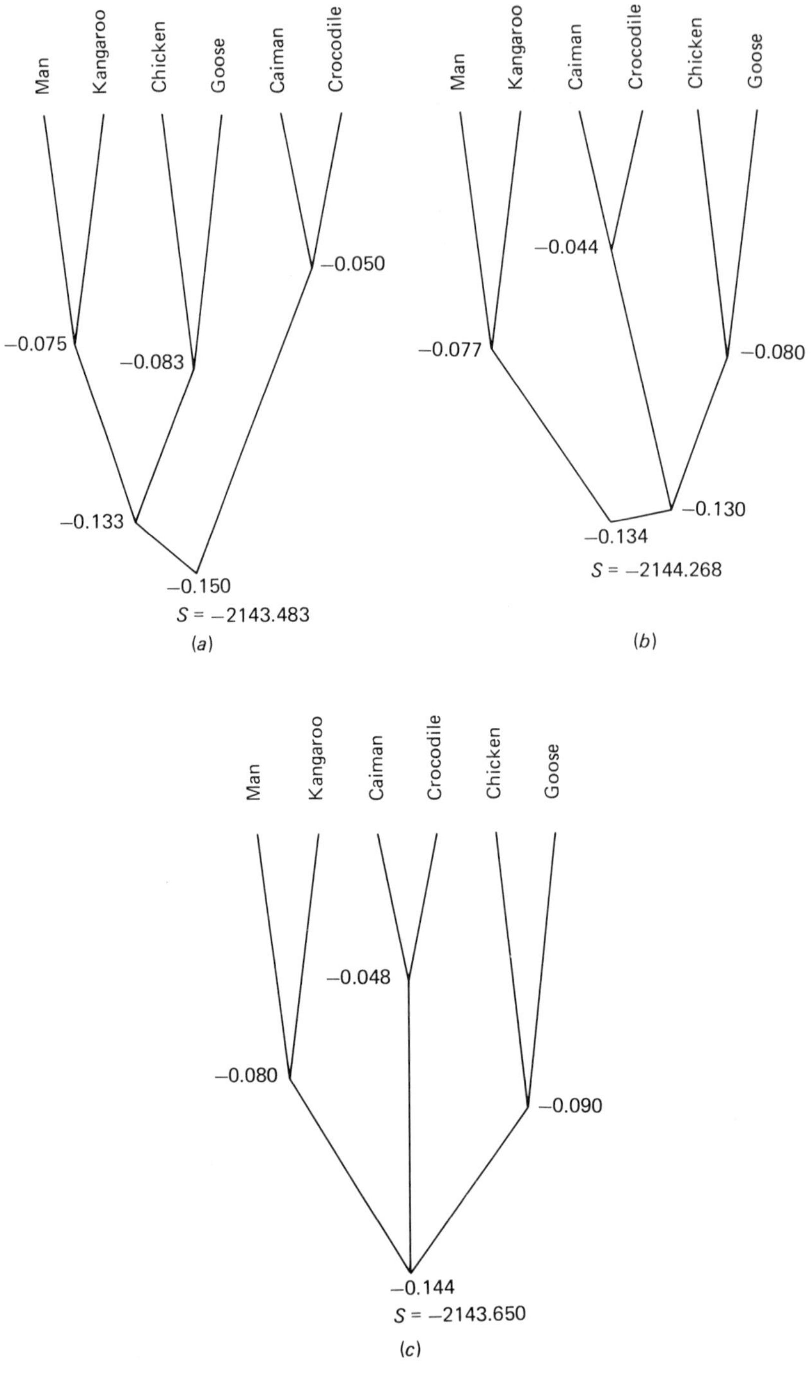
Man
Kangaroo
Chicken
Goose
Caiman
Crocodile
−0.050
−0.075
−0.083
−0.133
−0.150
S = −2143.483
(a)
Man
Kangaroo
Caiman
Crocodile
Chicken
Goose
−0.044
−0.077
−0.080
−0.130
−0.134
S = −2144.268
(b)
Man
Kangaroo
Caiman
Crocodile
Chicken
Goose
−0.048
−0.080
−0.090
−0.144
S = −2143.650
(c)

rates of homoiothermic mammals and birds have led to convergently acquired similarities in their oxygen-binding proteins. The morphological similarities, noted above, in the hearts and circulatory systems of the two groups could likewise have been convergently acquired in association with the high metabolic demands.

Perutz *et al.* (1981) also included the alpha haemoglobin of viper (*Vipera aspis*) in their study. They found that the viper joined their estimated tree as sister-group to the mammal plus bird plus crocodilian assemblage, and they drew attention to this unexpected placement. We have found the same placement for viper using pairwise maximum likelihood estimation. We can only reiterate the comments of Perutz *et al.* that, without further data, caution is required, but that the result suggests a divergence between archosaurian and lepidosaurian reptiles earlier than that between archosaurs and mammals. It is a pity that neither a chelonian nor a lizard sequence is yet available for alpha haemoglobin.

Conclusion

Our analysis seems to emphasize the similar problems encountered in the interpretation of biochemical and morphological data. This serves to underline the fact that both types of data need to be approached with equal caution.

REFERENCES

Benton, M. J. (1985). Classification and phylogeny of the diapsid reptiles. *Zoological Journal of the Linnean Society*, **84,** 97–164.

Bishop, M. J. & Friday, A. E. (1985). Evolutionary trees from nucleic acid and protein sequences. *Proceedings of the Royal Society, B*, **226,** 271–302.

Blanchetot, A., Wilson, V., Wood, D. & Jeffreys, A. J. (1983). The seal myoglobin gene, an unusually long globin gene. *Nature, London*, **301,** 732–4.

Deconinck, M., Peiffer, S., Depreter, J., Paul, C., Schnek, A. G. & Leonis, J. (1975). The primary structure of chicken myoglobin (*Gallus gallus*). *Biochimica et biophysica Acta*, **386,** 567–75.

Fig. 5.4 Maximum likelihood estimates of trees based on alpha-haemoglobin sequence data for six tetrapod species referred to in the text. The relative time estimates at the branching points, and the *S* values given for each tree, are described in the legend to Fig. 5.3. The three trees are little distinguished in terms of their associated likelihoods.

Dene, H., Sazy, J., Goodman, M. & Romero-Herrera, A. E. (1980). The amino acid sequence of alligator (*Alligator mississippiensis*) myoglobin. Phylogenetic implications. *Biochimica et biophysica Acta*, **624,** 397–408.

Felsenstein, J. (1981). Evolutionary trees from DNA sequences: a maximum likelihood approach. *Journal of Molecular Evolution*, **17,** 368–76.

Gaffney, E. S. (1979*a*). Tetrapod monophyly: a phylogenetic analysis. *Bulletin of the Carnegie Museum of Natural History*, **13,** 92–105.

Gaffney, E. S. (1979*b*). An introduction to the logic of phylogenetic reconstruction. In *Phylogenetic Analysis and Paleontology*, ed. J. Cracraft & N. Eldredge, pp. 79–111. New York: Columbia University Press.

Gaffney, E. S. (1980). Phylogenetic relationships of the major groups of amniotes. In *The Terrestrial Environment and the Origin of Land Vertebrates*, ed. A. L. Panchen, pp. 593–610. London: Academic Press.

Gardiner, B. G. (1982). Tetrapod classification. *Zoological Journal of the Linnean Society*, **74,** 207–32.

Holmes, E. B. (1975). A reconsideration of the phylogeny of the tetrapod heart. *Journal of Morphology*, **147,** 209–28.

Holmes, E. B. (1985). Are lungfishes the sister-group of tetrapods? *Biological Journal of the Linnean Society*, **25,** 379–97.

Holmgren, N. (1933). On the origin of the tetrapod limb. *Acta Zoologica, Stockholm*, **14,** 185–295.

Jarvik, E. (1942). On the structure of the snout of crossopterygians and lower gnathostomes in general. *Zoologiska Bidrag, Uppsala*, **21,** 235–675.

Jarvik, E. (1980). *Basic Structure and Evolution of Vertebrates*, vol. 2. London: Academic Press.

Litman, G. W., Berger, L., Murphy, K., Litman, R., Hinds, K., Jahn, C. L. & Erickson, B. W. (1983). Complete nucleotide sequence of an immunoglobulin V_H gene homologue from *Caiman*, a phylogenetically ancient reptile. *Nature, London*, **303,** 349–52.

Løvtrup, S. (1977). *The Phylogeny of Vertebrata*. London: John Wiley & Sons.

Løvtrup, S. (1986). On the classification of the taxon Tetrapoda. *Systematic Zoology*, **34,** 463–70.

Maeda, N. & Fitch, W. M. (1981*a*). Amino acid sequence of a myoglobin isolated from map turtle, *Graptemys geographica*. *Journal of Biological Chemistry*, **256,** 4293–300.

Maeda, N. & Fitch, W. M. (1981*b*). Amino acid sequence of a myoglobin from lace monitor lizard, *Varanus varius*, and its evolutionary implications. *Journal of Biological Chemistry*, **256,** 4301–9.

Owen, R. (1866). *On the Anatomy of Vertebrates*, vol. 2. London: Longmans, Green.

Patterson, C. (1980). Origin of tetrapods: historical introduction to the problem. In *The Terrestrial Environment and the Origin of Land Vertebrates*, ed. A. L. Panchen, pp. 159–75. London: Academic Press.

Peiffer, S. (1973). La myoglobine du manchot empereur (*Aptenodytes forsteri*). Thèse, Université libre de Bruxelles, Faculté des Sciences.

Perutz, M. F., Bauer, C., Gros, G., Leclercq, F., Vandecasserie, C., Schnek, A. G., Braunitzer, G., Friday, A. E. & Joysey, K. A. (1981). Allosteric regulation of crocodilian haemoglobin. *Nature, London*, **291,** 682–4.

Prass, W. A., Berkley, D. S. & Romero-Herrera, A. E. (1983). Chicken cardiac myoglobin revisited. *Biochimica et biophysica Acta*, **742,** 677–80.

Reig, O. A. (1967). Archosaurian reptiles: a new hypothesis on their origins. *Science*, **157**, 565–8.

Reisz, R. R. (1981). A diapsid reptile from the Pennsylvanian of Kansas. *Special Publications of the Museum of Natural History, University of Kansas*, **7**, 1–74.

Romer, A. S. (1971). Unorthodoxies in reptilian phylogeny. *Evolution*, **25**, 103–12.

Romero-Herrera, A. E., Lehmann, H., Joysey, K. A. & Friday, A. E. (1978). On the evolution of myoglobin. *Philosophical Transactions of the Royal Society, B*, **283**, 61–163.

Rosen, D. E., Forey, P. L., Gardiner, B. G. & Patterson, C. (1981). Lungfishes, tetrapods, paleontology, and plesiomorphy. *Bulletin of the American Museum of Natural History*, **167**, 159–276.

Thompson, E. A. (1975). *Human Evolutionary Trees*. Cambridge University Press.

Watson, D. M. S. (1926). The evolution and origin of the Amphibia. *Philosophical Transactions of the Royal Society, B*, **214**, 189–257.

Watts, D. A., Angelides, T. & Brown, W. D. (1983). The primary structure of myoglobin from Pacific green sea turtle (*Chelonia mydas caranigra*). *Biochimica et biophysica Acta*, **742**, 310–17.

Wells, R. M. G., Tetens, V. & Brittain, T. (1983). Absence of co-operative haemoglobin–oxygen binding in *Sphenodon*, a reptilian relict from the Triassic. *Nature, London*, **306**, 500–2.

Westoll, T. S. (1943). The origin of the tetrapods. *Biological Reviews*, **18**, 78–98.

Wiley, E. O. (1979). Ventral gill arch muscles and the interrelationships of gnathostomes, with a new classification of the Vertebrata. *Zoological Journal of the Linnean Society*, **67**, 149–79.

6 Pattern and process in vertebrate phylogeny revealed by coevolution of molecules and morphologies

Morris Goodman, Michael M. Miyamoto & John Czelusniak

Introduction

For purposes of this book we have examined major features of vertebrate phylogeny. Our mission has been to evaluate whether classical and molecular approaches give the same picture, or distorted views of the same picture, or two different pictures. With regard to the classical or morphological approach to vertebrate phylogeny, we have used such sources as Romer (1966), Løvtrup (1977) and Bonde (1975). With regard to the molecular approach, we have carried out by the maximum parsimony method genealogical reconstructions on substantial bodies of amino acid sequence data representing a wide range of vertebrates. As evident from the title of our paper, our verdict is that classical and molecular approaches yield basically the same picture or, rather, complementary views of the same phylogeny. Each approach sees something about phylogeny that the other misses but each approach also sees many of the same features.

The classical approach with its emphasis on the fossil record gives a time dimension to vertebrate phylogeny. The classical approach provides evidence on extinctions, speciations, and patterns of morphological change during the adaptive radiations that characterize vertebrate history. In this account we begin with a brief examination of the palaeontological picture in the time range of 475 to 340 million years before the present (MyrB P), and find that knowledge of the genealogy of major groups of vertebrates is still fragmentary and thus subject to improvement by the molecular approach. We then present the principles of molecular phylogenetics and outline our strategy in reconstructing vertebrate history from amino acid sequence data. Finally, we describe the vertebrate phylogeny emerging from this molecular approach, comparing it to the classical picture, first in terms of phylogenetic systematics – the cladistic relationships of major vertebrate branches – and secondly, in terms of tempos and modes of evolutionary change during different periods of vertebrate descent.

Early fossil history of vertebrates

The earliest remains surely indicative of vertebrate life are from early Ordovician sediments, about 475 MyrBP. These are fossilized dermal armour of primitive jawless vertebrates, the heterostracans. In the early Silurian, about 430 MyrBP, heterostracans are joined by other groups of jawless vertebrates, the anaspids, thelodonts and galeaspids. Vertebrates with jaws, the gnathostomes, are first recorded as early Silurian fragments, scales and spines from acanthodian fishes. By the late Silurian, close to its boundary with the Devonian at about 400 MyrBP, there is evidence of diverse jawless vertebrates, and among gnathostomes the acanthodians are joined by the first true bony fishes, scales and teeth which are suggestive of actinopterygians, the ray-finned fishes. In early Devonian sediments there is a wide range of jawless and jawed fishes (Fig. 6.1).

In these earliest vertebrate faunas virtually all the well-known forms are characterized by bizarre specializations; in other words, autapomorphic characters predominate, rather than the symplesiomorphic (shared primitive) characters that would be expected in any generalized ancestor. Nevertheless, among the early jawless vertebrates are two groups, the Osteostraci (cephalaspids) and Anaspida, which are generally regarded as genealogically related to Recent lampreys. The Heterostraci are a much more problematical group of jawless fishes: some have speculated that they are related to gnathostomes, others relate them to the Recent hagfishes, and others to cyclostomes (lampreys plus hagfishes) (see Janvier, 1981, and 1984 for review).

Early gnathostomes are about as puzzling as the early jawless fishes. The two dominant early groups (Fig. 6.1) are the acanthodians and placoderms, each characterized by obvious specializations, and each of disputed relationships. Some consider placoderms and acanthodians closest to Recent sharks and rays (e.g. Jarvik, 1980), others that both placoderms and acanthodians are related to osteichthyans (bony fishes), others that only one of the two is so related, and still others that either one group or the other is the sister-group of all gnathostomes. However, some Devonian fossils are less controversial, and all agree that lungfishes (Dipnoi), coelacanths (Actinistia), actinopterygians and chondrichthyans existed in the early Devonian, and that tetrapods appear in the late Devonian (Fig. 6.1). Thus, the fossil record shows a relatively rapid emergence of the major grades and clades of Vertebrata, and little or no solid evidence of intermediate forms. We can attribute these observations, as Darwin did, to the incompleteness of the fossil record; or we might take them as dramatic examples of punctuated equilibrium (Gould & Eldredge, 1977; Stanley, 1979), in which extremely rapid evolution characterizes the early stages of major radiations.

Disputes concerning genealogical relationships

Whether due to incompleteness of the fossil record or to rapid evolution leaving little or no trace of connecting links, it must be admitted that the classical approach has not resolved a number of questions concerning

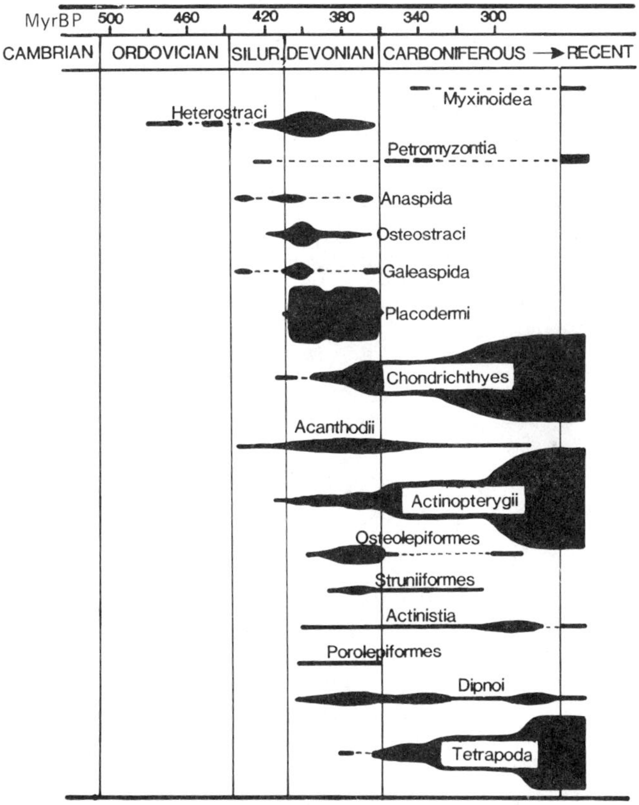

Fig. 6.1 The early fossil record of vertebrates. The chart shows the stratigraphic distribution of the major monophyletic groups of Palaeozoic vertebrates, with the approximate abundance of each (in terms of species recorded) indicated by the breadth of the shapes. Redrawn and modified from Janvier (1985) by exclusion of a few dubious records and inclusion of some subsequently published.

vertebrate phylogeny and systematics (Fig. 6.2). One difficulty for the uninitiated, such as we practitioners of molecular phylogenetics, is knowing when a taxon in traditional taxonomy is meant to represent simply a grade (e.g. Agnatha, Reptilia), or a clade as well as a grade (e.g. Gnathostomata, Mammalia). Some important issues in the phylogenetic classification of vertebrates which have not yet been satisfactorily answered by analyses of anatomical characters are the following.

First, are Cyclostomata – lampreys plus hagfishes – a poly- or paraphyletic group (i.e. an unnatural assemblage from the point of view of cladistic or genealogical classification), or a *bone fide* monophyletic taxon? Løvtrup (1977), Janvier (1981), Hardisty (1982) and Forey (1984) depict the former, placing lampreys genealogically closer to gnathostomes than to hagfishes, whereas the traditional view, supported by Romer (1966), Schaeffer & Thomson (1980) and Yalden (1985), depicts the latter, with lampreys and hagfishes as members of a true monophyletic lineage.

Second, the class Osteichthyes for bony fishes, as traditionally constituted, is certainly not a monophyletic clade, but rather a paraphyletic or even polyphyletic grade because tetrapods, which are not included in this class, are nevertheless thought to be derived from some member or members of the extinct group Rhipidistia. We can resolve this taxonomic problem in a Hennigian way simply by including tetrapods within Osteichthyes, and various classifications along these lines have been proposed (e.g. von Wahlert, 1968; Nelson, 1969; Wiley, 1979*b*; Rosen *et al.*, 1981).

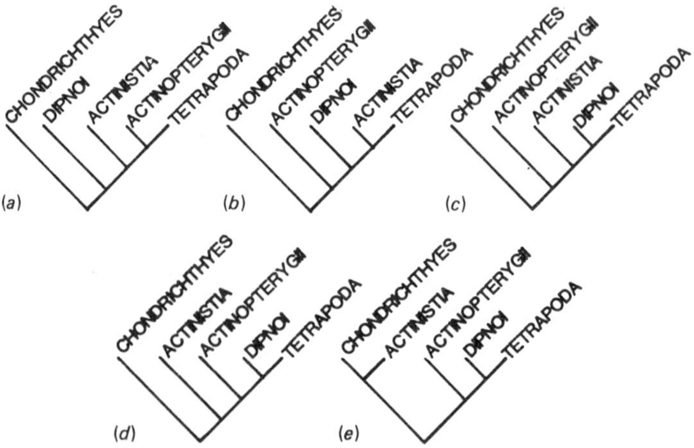

Fig. 6.2 Alternative phylogenies of the major groups of extant gnathostomes. The branching diagrams summarize the opinions of (*a*) von Wahlert (1968), (*b*) Romer (1966), (*c*) Rosen *et al.* (1981), (*d*) Wiley (1979b), and (*e*) Løvtrup (1977).

The question still remains, however, as to which surviving branch of bony fishes is the sister-group of tetrapods and, more generally, what are the genealogical relationships, or patterns of branching, among the major surviving groups, i.e. actinopterygians, tetrapods, coelacanths and lungfishes? The 'establishment' view, such as depicted by Romer (Fig. 6.2*b*), groups coelacanths with tetrapods, with the lungfishes joining next, and then the actinopterygians. An alternative view (e.g. Rosen *et al.*, 1981; Fig. 6.2*c*) exchanges the position of lungfishes and coelacanths, placing lungfishes as the sister-group of tetrapods. Wiley's (1979*b*) scheme (Fig. 6.2*d*) differs in placing coelacanths as the sister-group of all other osteichthyans. Yet another view, that of von Wahlert (1968; Fig. 6.2*a*), groups tetrapods with actinopterygians, with coelacanths joining next, and lungfishes as the sister-group of all other osteichthyans. In all these schemes Chondrichthyes are the sister-group of Osteichthyes, but even that is not beyond dispute, for Løvtrup (1977; Fig. 6.2*e*) groups coelacanths with Chondrichthyes, and Jarvik (1980) treats lungfishes as relatives of chondrichthyans.

Third, with regard to tetrapods, disputes exist as to whether the two main branches of surviving amphibians, Caudata (salamanders) and Anura (frogs), are sister-groups, constituting a monophyletic Amphibia (Jurgens, 1971), or whether Anura are closer to Amniota than to Caudata (see Løvtrup (1977) and Gardiner (1983) for reviews).

Inasmuch as the views of Romer (1966) and Løvtrup (1977) typify, within the purview of the classical morphological approach, disputes about the course of vertebrate phylogeny, these two views are contrasted in Fig. 6.3. It may be noted in this figure that the classical approach also fails to give an agreed picture of branching within Tetrapoda. For example, while Romer and Løvtrup agree on placing Crocodylia as the sister-group of Aves, in a monophyletic Archosauria, Romer places the archosaurs closest to squamates (lizards and snakes), whereas Løvtrup places archosaurs closest to testudinates (turtles). A minority view (Gardiner, 1982; Bishop & Friday, this volume, pp. 133–5) is that Aves are the sister-group of Mammalia, not Crocodylia.

Principles of molecular phylogenetics

Clearly, a number of outstanding questions concerning the course of vertebrate phylogeny have not been resolved by traditional palaeontological and comparative anatomical evidence. We suspect that the reason for this lack of resolution is that, in practice, when dealing with morphological characters, the distinction between homologous and analogous structures is often ambiguous. Yet this distinction must be drawn because only homologous parts can reveal the genealogical relationships of different species. In theory the distinction is that similar structures or parts in different species are homologous if they trace back to a single genealogical precursor, the common ancestral species, whereas similar parts are analogous if they result from convergent evolution such as may be caused by natural selection acting on organisms of different phyletic origins that have adopted similar ways of life. Take wings in bats, birds and butterflies: clearly, these three types of wing are analogous, but in birds and bats, as tetrapod forelimbs, they are also homologous. The distinction between homology and analogy is simple here, for we have a well-understood and uncontradicted system of relationships showing that butterflies and bats are related not to birds, but respectively to insects (originally wingless) and to mammals (wingless, but with forelimbs). Obviously, things are rarely so clear-cut, and we run into trouble when dealing with characters such as the long arms (relative to the legs) of orang utan, gorilla and chimpanzee. We want to know how the character 'long arms' bears on the relationships

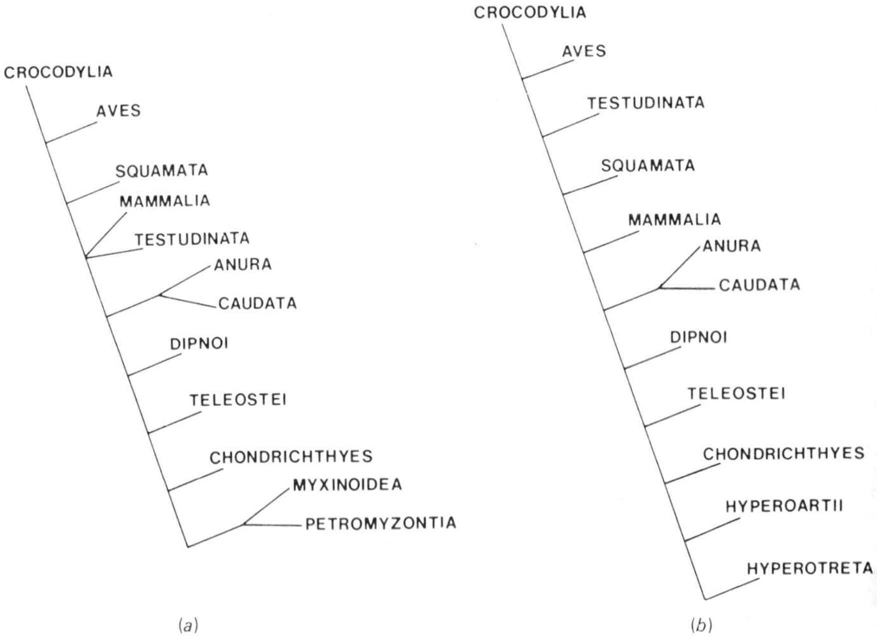

Fig. 6.3 Phylogenetic relationships of major vertebrate groups following (*a*) Romer (1966; see also Romer & Parsons, 1977) and (*b*) Løvtrup (1977). Romer's taxa Myxinoidea and Petromyzontia correspond to Løvtrup's groups Hyperotreta and Hyperoartii, respectively.

between these species, but we cannot decide whether the character is homologous or was acquired convergently until we have an accepted scheme of relationships between the great apes. In other words, possible homologies are evaluated by their congruence with a classification or phylogeny, which must itself be based on homologies.

A more basic problem with morphological characters as indicators of genealogical relationships is that there is no direct correspondence between the characters and heritable information encoded in genomic DNA. Subjectivism enters into not only the choice of characters but also the evaluation of whether a morphological difference between two species represents a large or small genetic difference. The big human brain, which in traditional taxonomy gives us a whole family to ourselves, may represent only a few small genic changes in the DNA of our lineage since we last shared a common ancestor with chimpanzees.

In contrast to the problems morphologists have, molecular biologists working with amino acid and nucleotide sequence data face very little ambiguity in deciding whether the sequences under examination are homologous. In comparing any two DNA nucleotide sequences, for instance, simple statistical procedures can reveal if the number of matching nucleotide bases is much greater than would be expected by chance, and so indicative of descent from a common genealogical precursor. (In terms of nucleotide sequences, there seems to be no equivalent of convergence, or close similarity produced by evolution from different precursors.) In comparing any two protein amino acid sequences, the minimum nucleotide differences can be calculated using the genetic code, and again, by statistical procedures, it is possible to decide whether the sequences are homologous. Computer programs exist to align any two sequences so as to maximize base matches, and to measure the distance between them (e.g. Smith & Waterman, 1981). Such distances are smaller for homologous sequences than for sequences showing only random matches (examples in Goodman *et al.*, 1984).

Once homologous protein or DNA sequences are identified in the species under study, they can be used to reconstruct the path of evolutionary descent of these species. This can be accomplished with a computer algorithm which constructs the evolutionary tree that accounts for the descent of the sequences by the fewest number of base changes (Moore, Barnabas & Goodman, 1973; Goodman *et al.*, 1979; Goodman, 1981). Such a maximum parsimony or maximum homology procedure (Moore, 1976) ensures that the Darwinian hypothesis on evolutionary descent is followed. In the Darwinian hypothesis genetic likeness results more often from shared common ancestry than from convergence. Every conceivable genealogical arrangement for a set of species may be entertained when no check is placed on the extent of convergence. When the genealogical reconstruction involves many homologous sequences

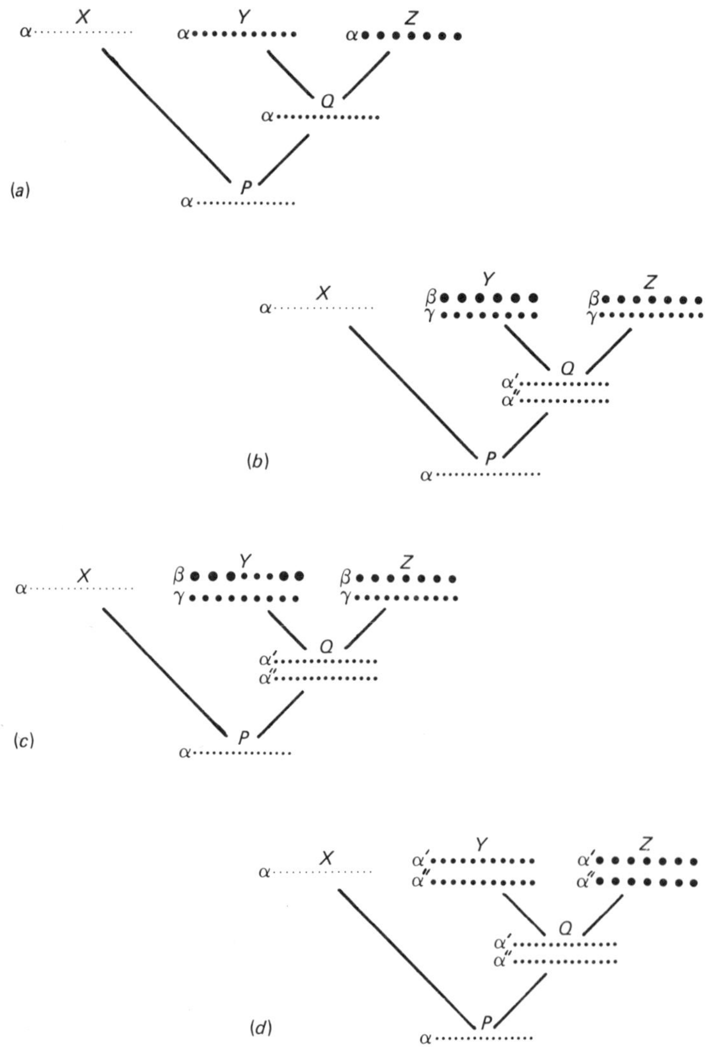

Fig. 6.4 Possible relations between molecular sequences, and their historical explanation. Sequences and their similarity are indicated by rows of dots. (*a*) Orthology: The sequences in species *X*, *Y* and *Z* are more similar to each other than to any other known sequence in the three taxa, and can be assumed to be strictly homologous, each the direct descendant of a sequence in ancestral species *P*. (*b*) Paralogy: Two sequences (beta and gamma) are found in species *Y* and in *Z*, but in *X* there is only one (alpha). Matching between the two betas is closer than between either and a gamma, and vice versa, whereas the alpha shows a general resemblance to the other four but does not match any particular one. The two betas are assumed to be orthologous, as are the two gammas, but within or between *Y* and *Z* the beta and gamma are

and thus many ancestral nodes to the reconstructed tree, the parsimony procedure – by inferring later substitutions superimposed over earlier ones – yields a larger and surely truer estimate of genic changes along the branches than do phenetic clustering algorithms that utilize only distance matrices from pairwise comparisons of the contemporary sequences.

However, to appreciate the fact that molecular biologists, in their efforts to reconstruct phylogeny from sequence data, encounter problems comparable to those faced by morphologists, we need only consider what a correctly constructed genealogical tree will reveal. It will reveal not only events of speciation but also events of gene duplication which have occurred during molecular evolution. To do so, this correct genealogical tree distinguishes between two types of homologous sequences when both are present, *orthologously* related sequences versus *paralogously* related sequences. As first distinguished and defined by Fitch (1970), *orthologous* sequences (Fig. 6.4*a*) are descendants of gene-lineages whose splittings from one another coincide with the splittings of the species-lineages within which they occur; *paralogous* sequences (Fig. 6.4*b*) are descendants of gene-lineages whose splittings occurred by gene duplication prior to speciation of the species-lineages. Events of gene conversion, whereby portions of a duplicated gene are replaced by paralogous portions of the other gene, may further complicate the comparison of homologous sequences among species, in that some portions of such sequences will be orthologously related and other portions paralogously related (Fig. 6.4*c*). Thus, ideally, a correctly constructed genealogical tree describing the evolution of a family of homologous genes should reveal not only paralogies due to gene duplication but also those paralogies due to later gene conversions. Such a correctly constructed gene phylogeny might also reveal orthologies due to gene conversion (Fig. 6.4*d*). For example, humans and many other primates have in each of their genomes two non-allelic loci that encode the α haemoglobin chain, suggesting that an ancient

Caption for Fig. 6.4 (*contd*)

paralogous, assumed to be descendants of an alpha-like gene, duplicated at some time between ancestral species *P* and *Q*. (*c*) Sequences part orthologous, part paralogous: Differs from (*b*) in that a portion (or portions) of one of the two sequences in species *Y* matches the corresponding (paralogous) part of the other sequence in that species, whereas the remainder matches the orthologous sequence in species *Z*. Gene conversion is assumed to have occurred in lineage *Y*, replacing part of a gene by its paralogue. (*d*) Sequences paralogous, but masquerading as orthologous: Differs from (*b*) in that the two sequences in species *Y* are virtually or completely identical, as are the two in species *Z*. Gene conversion is assumed to have occurred in lineages *Y* and *Z*, preventing or overriding the divergence between the two paralogous sequences inherited from ancestral species *Q*.

duplication of the α globin gene occurred, yet each such species has two α loci that encode identical α haemoglobin sequences; the sequence variation observed is not within species but between species, presumably because of gene conversions within the lines of species descent.

In reconstructing the phylogeny of a set of homologous sequences, convergent base substitutions unequally distributed among the different homologous sequences are the principal source of error. Such unequal homoplasy can cause mistakes of grouping in that sequences with excesses of convergent residues will be depicted as sharing a more recent common ancestor than was in fact the case. The way to get around this problem, when using sequence data to reconstruct the phylogeny of species, is to extend the alignment to a much larger number of sequence positions. The extended alignment increases the pool of variable positions and thereby lessens the chances of an unequal distribution of homoplasy among the contemporary species represented by the sequences. The principle is the same as distinguishing homology from analogy (= homoplasy, non-homology) in the wings of bats, birds and butterflies: the wings alone might mislead, but if one also compares the morphology of, say, the gut, brain and jaws in the three, the mistake will be revealed.

In the present study, as described in the next section, we have taken single sets of homologous protein sequences and combined them into an extended tandem alignment as if each species had a giant gene locus encoding the amino acid sequence of an extremely long polypeptide chain. From such tandemly combined sequences, a phylogeny of mammals (Miyamoto & Goodman, 1986) and a phylogeny of vertebrates as a whole were constructed by the maximum parsimony procedure, the trees describing these species phylogenies being those of lowest nucleotide replacement (NR) length. For the purpose of constructing the species phylogeny, the tandemly arranged data sets should have only orthologous sequences aligned against one another. In practice we followed a loose definition of orthology: as long as we found no evidence against the hypothesis that the homologous sequences diverged ancestrally from one another close in time to the ancestral divergence of their species lineages, they were treated as orthologues even when the overall evidence suggested that certain of the sequences included were paralogous to some of the other sequences. For example, in the phylogeny of globins that was constructed (in another phase of the present study) for a data set of 218 homologous globin sequences, the teleost β haemoglobin branch grouped with the tadpole β branch and this teleost–tadpole β branch then grouped with the large branch containing all the adult tetrapod βs (within which was the anuran β clade). Thus, a gene duplication preceding the ancestral split of teleosts and tetrapods has to be postulated; however, as will be apparent later when we present the genealogical tree of globin sequences, this gene duplication appears to have occurred close to the geological time when the last common ancestor of teleosts and tetrapods existed. Therefore, by our loose definition of orthology, we could still have teleosts represented in the region of the tandem alignment reserved for β sequences.

Genealogical reconstruction strategy

Table 6.1, the Appendix and Figs 6.5 and 6.6 present information on the taxa and corresponding protein sequences used in the present study. While the vast majority of our 117 taxa belong to the mammalian infraclass Eutheria (72), an appreciable number (45) belong to other major branches of Vertebrata. As is apparent in Fig. 6.5 and Table 6.1, the full range of vertebrates is best represented by globins and next best by lens α crystallins and cytochrome *c*. Within Gnathostomata, the globins subdivide into myoglobins, α-type haemoglobins and β-type haemoglobins, each type having orthologues that encompass many of the same vertebrate taxa. We considered including additional proteins in the extended tandem alignments, but could not identify any representing a sufficient range of vertebrates to merit inclusion.

Our strategy took advantage of a study that we carried out (Miyamoto & Goodman, 1986) on the phylogenetic systematics of eutherian mammals. This study focused on 90 operational taxonomic units (OTUs) of which 72 OTUs were from 16 of the 18 orders of extant eutherians; Dermoptera (flying lemur) and Macroscelidea (elephant shrews) could not be included because of lack of amino acid sequence data on these two minor eutherian orders. The sequences of the extended tandem alignment were presumed orthologues of myoglobin, α and β chains of typical adult haemoglobin, lens α crystallin A, cytochrome *c*, fibrinopeptides A and B, and ribonuclease. One rule generally followed was to include in the alignment as many sequences for the seven proteins as possible by erecting for eutherians hybrid OTUs, each such OTU representing a supraspecific taxon which has sequences from different species. For example, the OTU Atelinae (a New World monkey subfamily) consists of *Lagothrix lagothricha* (woolly monkey) myoglobin and *Ateles geoffroyi* (spider monkey) α and β haemoglobins, fibrinopeptides A and B, and cytochrome *c*.

The data set of 90 OTUs contained many more eutherian taxa and a denser and more extensive body of aligned sequences than in prior analyses from our laboratory (Goodman, 1981; Goodman *et al.*, 1982*a*; Goodman, Weiss & Czelusniak, 1982*b*; Goodman, Czelusniak & Beeber, 1985). A consequence of so increasing the sequence data on eutherians was that the trees of lowest NR length, found by our maximum parsimony algorithm on examining thousands upon thousands of alternative dendrograms, were in

Table 6.1. *Amino acid sequences for vertebrate OTUs used in the tandem alignment*

The vertebrate phylogeny (Fig. 6.7) is based on these data. OTUs reflecting supraspecific taxa constitute hybrids formed from combinations of the accompanying, indented species. Supraspecific groups follow the traditional taxonomy of Simpson (1945), Romer (1966), Løvtrup (1977), and Romer & Parsons (1977). Use of the taxa Palaeognathiformes and Polycryptodira follows Cracraft (1981) and Gaffney (1984), respectively. Abbreviations include: αHB, α haemoglobin; βHB, β haemoglobin; MB, myoglobin; LENS, lens α crystallin A; CYT, cytochrome *c*; and Eα, embryonic α haemoglobin.

Class	Order	OTU	Number of sequences	αHB	βHB	MB	LENS	CYT	Eα
Mammalia	Artiodactyla	*Bos taurus*	5	+	+	+	+	+	–
		Sus scrofa	6	+	+	+	+	+	+
		Subfamily Camelinae							
		Camelus dromedarius	4	+	+	–	+	+	–
		Lama vicugna	1	–	–	+	–	–	–
	Carnivora	*Canis familiaris*	5	+	+	+	+	+	–
		Family Phocidae							
		Halichoerus grypus	1	–	–	–	+	–	–
		Mirounga leonina	1	–	–	–	–	+	–
		Phoca vitulina	3	+	+	+	–	–	–
	Cetacea	Suborder Mysticeti							
		Balaenoptera acutorostrata	4	+	+	+	+	–	–
		Eschrichtius glaucus	1	–	–	–	–	+	–
	Chiroptera	Order Chiroptera							
		Artibeus jamaicensis	1	–	–	–	+	–	–
		Miniopterus schreibersi	1	–	–	–	–	+	–
		Rousettus aegyptiacus	3	+	+	+	–	–	–

Class	Order	OTU	Number of sequences	αHB	βHB	MB	LENS	CYT	Eα
	Marsupialia	Genus *Macropus*							
		Macropus cangura	1	−	−	−	−	+	−
		Macropus giganteus	2	+	+	−	−	−	−
		Macropus rufus	2	−	−	+	+	−	−
	Perissodactyla	*Equus caballus*	5	+	+	+	+	+	−
	Rodentia	Suborder Hystricomorpha							
		Cavia porcellus	4	+	+	−	+	+	−
		Lagostomus maximus	1	−	−	+	−	−	−
		Superfamily Muroidea							
		Rattus norvegicus	4	+	+	−	+	+	−
		Spalax ehrenbergi	1	−	−	+	−	−	−
	Lagomorpha	*Oryctolagus cuniculus*	5	+	+	+	+	+	−
	Primates	*Homo sapiens*	6	+	+	+	+	+	+
		Family Cebidae							
		Aotus trivirgatus	1	−	−	−	+	−	−
		Ateles geoffroyi	3	+	+	−	−	+	−
		Lagothrix lagothricha	1	−	−	+	−	−	−
		Family Lorisidae							
		Galago crassicaudatus	1	−	−	−	+	−	−
		Nycticebus coucang	4	+	+	+	−	+	−
		Genus *Macaca*							
		Macaca fascicularis	1	−	−	+	−	−	−
		Macaca mulatta	4	+	+	−	+	+	−
Aves	Anseriformes	*Anas platyrhynchos*	4	+	+	−	+	+	−
		Anser anser	2	+	+	−	−	−	−
		Anser indicus	2	+	+	−	−	−	−
		Anseranas semipalmata	2	+	+	−	−	−	−
		Branta canadensis	2	+	+	−	−	−	−
		Cairinia moschata	3	+	+	−	−	−	+
		Cygnus olor	3	+	+	−	+	−	−

Table 6.1. (*contd*)

Class	Order	OTU	Number of sequences	αHB	βHB	MB	LENS	CYT	Eα
	Ciconiiformes	*Ciconia ciconia*	2	+	+	–	–	–	–
		Phoenicopterus ruber	2	+	+	–	–	–	–
	Columbiformes	*Columba livia*	2	–	–	–	+	+	–
	Falconiformes	Family Accipitridae							
		Aquila chrysaetos	2	+	+	–	–	–	–
		Buteo buteo	1	–	–	–	+	–	–
	Galliformes	*Gallus gallus*	6	+	+	+	+	+	+
		Meleagris gallopavo	2	–	–	–	+	+	–
		Phasianus colchicus	2	+	+	–	–	–	–
	Palaeognathiformes	*Dromaius novaehollandiae*	2	–	–	–	+	+	–
		Rhea americana	3	+	+	–	+	–	–
		Struthio camelus	4	+	+	–	+	+	–
	Passeriformes	Order Passeriformes							
		Corvus corone	1	–	–	–	+	–	–
		Sturnus vulgaris	2	+	+	–	–	–	–
	Sphenisciformes	Order Sphenisciformes							
		Aptenodytes forsteri	1	–	–	+	–	–	–
		Aptenodytes patagonica	1	–	–	–	–	+	–
		Pygoscelis papua	1	–	–	–	+	–	–
Reptilia	Crocodylia	*Alligator mississippiensis*	4	+	+	+	+	–	–
		Caiman latirostris	2	+	+	–	–	–	–
		Crocodylus niloticus	2	+	+	–	–	–	–
	Squamata	Suborder Sauria							
		Tupinambis teguixin	1	–	–	–	+	–	–
		Varanus varanus	2	–	–	+	–	+	–
		Suborder Serpentes							
		Crotalus adamanteus	1	–	–	–	–	+	–
		Vipera aspis	1	+	–	–	–	–	–

Table 6.1. (*contd*)

Class	Order	OTU	Number of sequences	αHB	βHB	MB	LENS	CYT	Eα
	Testudinata	Suborder Polycryptodira							
		Chelonia mydas	1	−	−	+	−	−	−
		Chelydra serpentina	1	−	−	−	−	+	−
Amphibia	Anura	*Xenopus laevis*	3	+	+	−	−	−	+
		Genus *Rana*							
		Rana catesbeiana	3	−	+	−	−	+	+
		Rana esculenta	1	−	−	−	+	−	−
	Caudata	*Ambystoma mexicanum*	1	+	−	−	−	−	−
		Taricha granulosa	1	+	−	−	−	−	−
Osteichthyes	Cypriniformes	*Carassius auratus*	2	−	+	−	−	−	+
		Catostomus clarkii	1	−	−	−	−	−	+
		Cyprinus carpio	4	−	+	+	−	+	+
	Dipnoi	*Lepidosiren paradoxus*	3	+	+	−	−	−	+
	Salmoniformes	*Salmo irideus*	1	−	−	−	−	−	+
	Perciformes	*Katsuwonus vagrans*	1	−	−	−	−	+	−
		Genus *Thunnus*							
		Thunnus albacares	1	−	−	+	−	−	−
		Thunnus thynnus	1	−	−	−	−	+	−

Table 6.1. (*contd*)

Class	Order	OTU	Number of sequences	αHB	βHB	MB	LENS	CYT	Eα
Chondrichthyes	Selachii	Order Selachii							
		Heterodontus portusjacksoni	4	+	+	+	−	−	+
		Squalus acanthias	1	−	−	−	+	−	−
		Squalus sucklii	1	−	−	−	−	+	−
Agnatha	Myxinoidea	*Myxine glutinosa*	4	+	+	+	−	−	+
	Petromyzontia	Order Petromyzontia							
		Entosphenus tridentatus	1	−	−	−	−	+	−
		Lampetra fluviatilis	4	+	+	+	−	−	+
Arthropoda	Diptera	Order Diptera							
		Chironomus thummi	4	+	+	+	−	−	+
		Drosophila melanogaster	1	−	−	−	−	+	−
Gastropoda	—	Class Gastropoda							
		Aplysia limacina	4	+	+	+	−	−	+
		Helix aspersa	1	−	−	−	−	+	−
Totals			197	44	44	28	31	34	16

almost all respects congruent with the evidence on eutherian phylogeny based on traditional morphological characters. The only non-congruent feature with traditional evidence is that the new trees of lowest NR length for tandemly combined sets of amino acid sequences still depict Monotremata rather than Marsupialia as the sister-group of Eutheria. However, it should be pointed out that the sequence data on monotremes and marsupials have not increased since our earlier studies, and monotremes are still represented solely by the three globins (myoglobin and α and β haemoglobins). Moreover, as will be apparent when we describe the genealogical reconstruction carried out on 218 globin sequences (Fig. 6.5), α haemoglobin sequences do depict the traditional arrangement, of Prototheria (monotremes) as the sister-group of Theria (marsupials and eutherians); it is just the myoglobin and β haemoglobin sequences which depict monotremes as an ancient branch of Eutheria and are therefore incongruent with traditional characters.

We anticipate that eventually, after more extensive sequence data are obtained on monotremes, the incongruent position of Monotremata will be resolved by the most parsimonious trees constructed (see also McKenna, this volume, p. 59). An analogous problem existed for Crocodylia and Aves with regard to their respective positions to each other and Mammalia (see also Bishop & Friday, this volume, p. 133). In our reconstructed phylogeny of globins (Fig. 6.5) only α haemoglobin sequences are congruent with traditional morphological evidence by depicting a monophyletic Archosauria with Crocodylia the sister-group of Aves; both myoglobin and β haemoglobin sequences move Mammalia ahead of crocodilians and thereby depict Mammalia as the sister-group of Aves. However, as will be seen in the most parsimonious tree (Fig. 6.7) constructed for an extended tandem alignment of the different protein sequences representing a broad range of vertebrates (Table 6.1), addition of cytochrome *c* and lens α crystallin A sequences to the analysis results in the overall molecular evidence being congruent with morphological evidence: among extant vertebrates Crocodylia are the sister-group of Aves, the two clades forming a monophyletic Archosauria.

We can now turn to how our protein sequence results on eutherian cladistics (Miyamoto & Goodman, 1986) shaped our strategy for uncovering the cladistic pattern of major vertebrate branches. At the time we began this study, the available body of globin amino acid sequences (from over 400 different globin chains) so exceeded the parameters of our computer programs that we did not attempt to include all these sequences in one common alignment. Instead, representation of Eutheria was restricted by requiring that every included eutherian species be represented by all three types of globin sequences, myoglobin and typical α- and β-type chains. There were 35 such species. This choice allowed us to have each set of eutherian orthologues follow the same branching arrangement (Fig. 6.6). This arrangement was the one of lowest NR

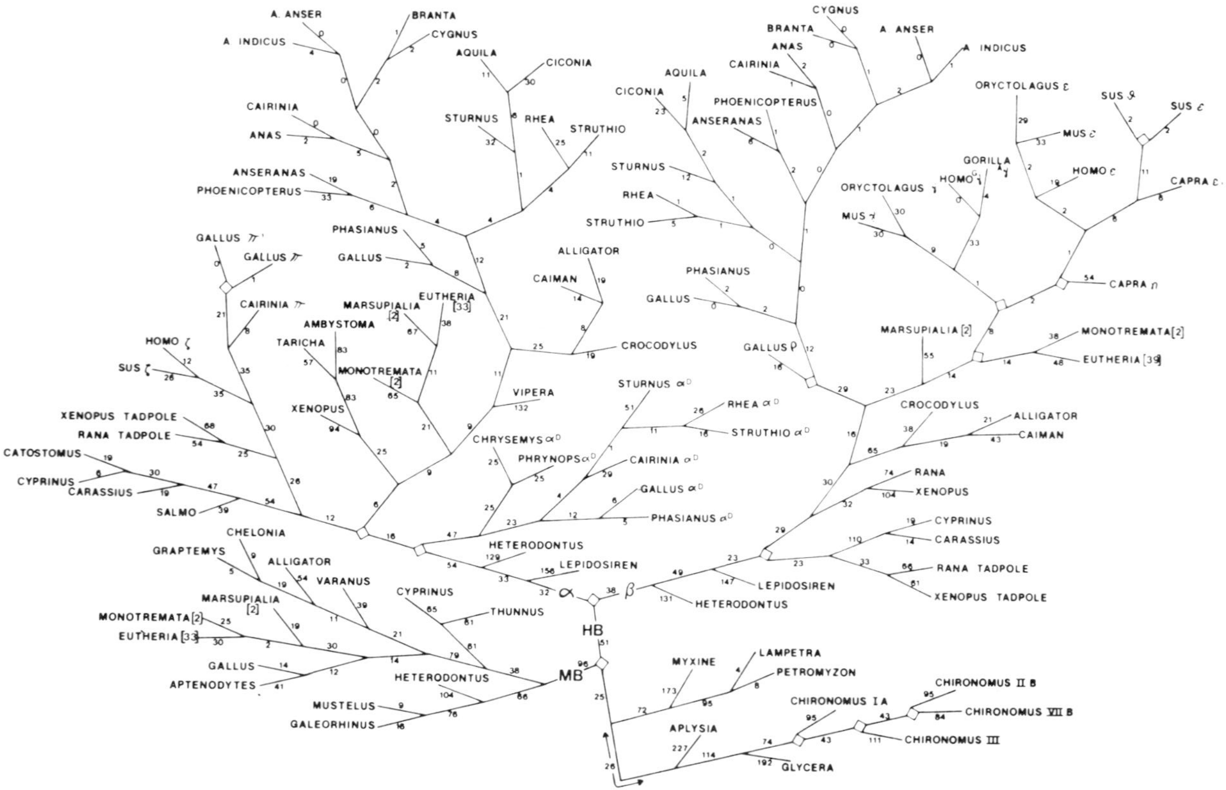

Fig. 6.5 Globin phylogeny for 218 sequences. This phylogeny requires 4813 nucleotide replacements. Link lengths, augmented for superimposed mutations (see Goodman, 1981), are presented next to branches. Diamonds refer to obvious gene duplications. The eutherian regions of the tree (not shown) conform to the branching arrangement found for these taxa by Miyamoto & Goodman (1986). Numbers in square brackets for eutherians, marsupials and monotremes correspond to the numbers of haemoglobin and myoglobin sequences used in the analysis. Most sequences used here are catalogued in Goodman (1981) and Goodman *et al.* (1982*a*, *b*, 1985). Those not previously cited are listed in the Appendix.

Polychotomies are represented in the tree by dichotomies associated with zero branch lengths. Although not directly supported, length found with tandemly combined sets of protein sequences in our study of eutherian cladistics. We further saved computing time by including only a few non-vertebrate globins and by employing the sequence alignment of non-vertebrate globins against vertebrate globins previously found (Goodman, 1981).

The fact that the same species-lineages reoccurred in different branches of the globin phylogeny obviously imposed a constraint on our search for the most parsimonious arrangement of sequences within these branches, that the different sets of orthologues should depict the same species relationships. While generally following this constraint, we did not hold to it whenever there was strong reason to suspect that certain of the presumed orthologues might be paralogues or when no clear trend emerged as to the exact cladistic position of a species-lineage represented by different globins.

The most parsimonious tree found for the 218 globin sequences (Fig. 6.5), upon examining thousands of dendrograms, revealed that, in addition to myoglobin and typical α and β chains of adult haemoglobin, an embryonic type of α chain (represented by teleost, tadpole, avian and mammalian sequences) provided evidence on cladistic relationships of major vertebrate groups. All teleost and tetrapod α sequences, whether of embryonic or adult type, were found, as shown in Fig. 6.5, to be cladistically closer to one another than to lungfish (*Lepidosiren*) and shark (*Heterodontus*) α sequences. A similar finding was made for β sequences in which, again, teleosts appear as the sister-group of tetrapods. With regard to the most ancient branches of this globin phylogeny, monophyletic α and β branches of gnathostomes join each other and then are joined by a monophyletic gnathostome myoglobin (MB) branch; the three major gnathostome branches are cladistically closer to one another than to cyclostome, mollusc, worm and dipteran globins. These various findings were utilized in combining the different sets of sequences in an extended tandem alignment for the purpose of constructing by the maximum parsimony method a phylogeny of vertebrates. As recorded in Table 6.1, this extended alignment consists of myoglobin, β globin, adult type α globin, embryonic type α globin, lens α crystallin A and cytochrome *c*. Teleost αs were placed in the embryonic α region of the alignment. Lungfish α and shark α were each placed in both adult and embryonic α regions. Each globin from cyclostomes and invertebrates was placed in all four globin regions of the alignment.

The 57 OTUs consisted of 16 therians, 19 birds, 20 other vertebrates, one gastropod and one dipteran. The criteria followed in choosing these OTUs were as follows:

1 Each mammal had to be represented by the three adult type globins, lens α crystallin A, and cytochrome *c*.
2 Each bird had to be represented by two or more protein chains.

3 Each non-mammalian and non-avian vertebrate needed no more than one protein chain to be included.
4 Each non-vertebrate had to be a metazoan and be represented by the two kinds of proteins, globin and cytochrome *c*, that have been sequenced in non-vertebrate and vertebrate metazoans.

Among vertebrates, OTUs that were supraspecific hybrids were formed between members of an order whenever the taxa of each such hybrid failed to share a sequence in common. Among invertebrates, supraspecific hybrids were formed between members of a class when these members failed to share a sequence in common. We felt that at the class level and below (e.g. Gastropoda, Diptera), monophyletic OTUs for invertebrate metazoans could be formed, but that above the class level, the danger existed that a hybrid OTU might represent a polyphyletic group. In the search for the most parsimonious tree for these 57 metazoan OTUs, the branching arrangement of the 16

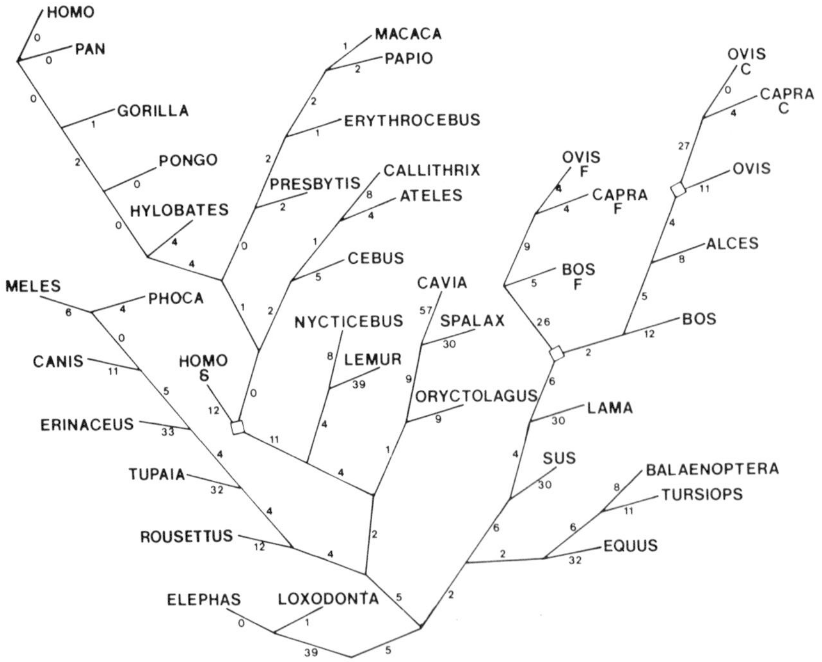

Fig. 6.6 Eutherian region of the β-branch in the overall globin phylogeny (Fig. 6.2). The species relationships among these eutherians follow the phylogeny of Miyamoto & Goodman (1986). The branching arrangements for these species in the α-haemoglobin and myoglobin regions of the globin phylogeny are identical to those shown here. Augmented link lengths are included next to branches (see Fig. 6.5). Diamonds refer to obvious gene duplications, such as those involving the additional paralogous loci ('F' and 'C') of *Bos*, *Capra* and/or *Ovis*.

therian OTUs was fixed according to the cladistic pattern found for mammalian OTUs in our study of eutherian cladistics (Miyamoto & Goodman, 1986). Otherwise, any change in branching patterns that lowered the NR length of this vertebrate phylogeny was acceptable. Under the conditions mentioned, the tree shown in Fig. 6.7 – found after examining many thousands of alternative dendrograms – had a lower NR length than any other tree examined during the search.

Molecular picture of vertebrate cladistics

Figure 6.7 serves as a synthesis of our molecular evidence on phylogenetic relationships among the major branches of extant vertebrates. The relationships depicted in this figure incorporate, as described in the preceding section, our results on eutherian cladistics (Miyamoto & Goodman, 1986), on globin phylogeny (Fig. 6.5), and on all those non-mammalian and non-avian vertebrates represented by at least one protein chain among globins, lens α-crystallin A, and cytochrome *c* (Table 6.1). A more explicit statement of this synthesis may be given in the form of a taxonomic classification in which each taxon is thought to represent a monophyletic group. The provisional classification shown in Table 6.2 represents such an effort. With one exception it includes only those vertebrate branches found in Fig. 6.7. The one exception concerns Coelacanthini (*Latimeria chalumnae*) which is represented by α and β parvalbumins in the fuller available body of amino acid sequence data. The most parsimonious trees constructed for homologues of these parvalbumin sequences have teleosts joined to tetrapods, with *Latimera* as their sister-group (Baba *et al.*, 1984; Maeda, Zhu & Fitch, 1984). On the basis of this parvalbumin evidence, we also list Coelacanthini as well as Dipnoi, Teleostei and Tetrapoda under Osteichthyes. Our classification is provisional, first because it is missing certain extant groups of cold-blooded vertebrates (those not yet represented by protein sequences), and second because certain relationships depicted may have to be revised after a denser and more extensive body of sequence data becomes available for cladistic analysis.

Using the method of phyletic sequencing (Wiley, 1979*a*, 1981), we describe for major vertebrate branches these same relationships in Table 6.3. This molecular picture of vertebrate cladistics agrees closely with the overall classical picture. As discussed in an earlier section of our paper, within the purview of the classical approach, there are a number of unresolved questions and disputes concerning the cladistic pattern revealed by morphology. With regard to these disputes, our molecular evidence favours certain views over others. In agreement with Romer (1966), Schaeffer & Thomson (1980) and Yalden (1985), our molecular evidence depicts a monophyletic Cyclostomata.

Fig. 6.7 Most parsimonious phylogeny for vertebrates based on amino acid sequences (Table 6.1). This tree requires 5024 nucleotide replacements. Numbers refer to the total sequences associated with each OTU. References for sequences used here but not cited in Goodman (1981) or Goodman *et al.* (1982*a*, *b*, 1985) are listed in the Appendix.

In addition to this tree, other solutions requiring 5024 nucleotide replacements are supported by the data. In most cases, these alternatives involve only minor rearrangements within and among orders of birds and teleosts. However, the basal position of the order Galliformes within the avian lineage disagrees with the traditional view (Cracraft, 1981). Furthermore, the relationships of caudate amphibians to amniotes, anurans and teleosts remain somewhat unclear, as additional hypotheses are supported by the tandem alignment. Nevertheless, as suggested by the tree shown here, caudates (represented only by α-haemoglobins) are best treated as the sister-group of anurans, following the results of the denser globin analysis based on a larger body of such data (Fig. 6.5). In the globin phylogeny, caudates strongly join the anuran *Xenopus* given available data (i.e. their α-haemoglobins), and, for now, we therefore regard them as sister taxa.

Table 6.2. *Provisional classification of vertebrates including only those monophyletic groups for which sequence data are available*

- Vertebrata
  - Agnatha (Cyclostomata)
    - Petromyzontia
    - Myxinoidea
  - Gnathostomata
    - Chondrichthyes (Elasmobranchii)
      - Heterodontoidea
      - Galeoidea
    - Osteichthyes
      - Dipnoi
      - Coelacanthini
      - Euosteichthyes
        - Actinopterygii (Teleostei)
          - Cypriniformes
          - Perciformes
          - Salmoniformes
        - Tetrapoda
          - Amphibia
            - Anura
            - Caudata
          - Amniota
            - Mammalia
            - Sauropsida
              - Testudinata
              - Squamata
              - Archosauria
                - Aves
                - Crocodylia
                  - Crocodylinae
                  - Alligatorinae

The molecular picture of Gnathostomata is identical in broad outline to von Wahlert's (1968); within Osteichthyes, Actinopterygii (Teleostei) – rather than either Dipnoi or Coelacanthini – are the sister-group of Tetrapoda. Concerning disputes on Tetrapoda, the molecular picture favours a monophyletic Amphibia, and within Amniota a division into Mammalia and Sauropsida. Within the latter, a grouping of squamates (lizards and snakes) plus turtles is the sister-group of Archosauria (crocodilians and birds).

Molecular picture of Darwinian evolution of vertebrates

The present analysis of globin phylogeny in the vertebrates (Fig. 6.5) agrees with previous analyses (Goodman, Moore & Matsuda, 1975;

Table 6.3. *Phylogenetic classification of vertebrates based on the amino acid phylogeny of Fig. 6.7. This classification summarizes the overall branching arrangement of the phylogeny using the conventions of subordination (ranking), phyletic sequencing, and annotation (Wiley, 1979*a*, *1981). In phyletic sequencing, taxa reflecting successive dichotomous branchings are assigned equivalent rank and are listed in the classification in their postulated order of descent, so that each is the sister-group of all listed beneath it. In contrast, polychotomies are represented in the classification by the annotation* sedis mutabilis *(L. – of changeable position)*

- Subphylum Vertebrata
  - Infraphylum Agnatha
  - Infraphylum Chondrichthyes
  - Infraphylum Osteichthyes (all superclasses *sedis mutabilis*)
    - Superclass Actinistia
    - Superclass Dipnoi
    - Superclass Euosteichthyes
      - Class Actinopterygii
      - Class Amphibia
      - Class Mammalia
      - Class Reptilia
      - Class Aves
      - Class Crocodylia

Goodman, 1981; Goodman *et al.*, 1982*a, b*). We again find extremely fast rates of globin evolution (Table 6.4) during the first great radiations of vertebrates. Even using Løvtrup's (1977) mid-Cambrian date of 535 MyrBP for the most ancestral vertebrate node (which is long before the appearance of vertebrate fossils, Fig. 6.1), we find for the stretches of time to the gnathostome and amniote ancestral nodes (at about 420 MyrBP and 310 MyrBP, respectively) almost an order of magnitude faster rates than in the amniote lineages descending to mammals and birds. After a marked slow-down in rates of globin evolution between 310 and 85 MyrBP in descent to the eutherian ancestral node, our phylogenetic reconstruction shows that rates speed up again during emergence and diversification of the orders of eutherian mammals. However, later in certain lineages, such as the one to humans, rates once more slowed drastically. A range of other proteins, including cytochrome *c*, lens α crystallin A, calmodulin, troponin C, regulatory light chains of myosin, and parvalbumins, also evolved at much faster rates during emergence of jawed vertebrates and basal tetrapods than during descent of Aves and Mammalia (Goodman, 1981; Baba *et al.*, 1984). Cytochrome *c* and lens α crystallin A, like globins, experienced accelerated evolution during the emergence of Eutheria and then decelerated evolution in later hominoids.

The pattern of speed-ups and slow-downs in rates, with a trend over time

Table 6.4. *Evolutionary rates for haemoglobins (HB) and myoglobins. These rates are based on the link length estimates of Figs 6.5 and 6.6, augmented for superimposed mutations, as described in Goodman (1981). NR% refers to the number of augmented nucleotide replacements per 100 codons per 100 million years. Estimates for the last eight periods are based on averages of all lineages descending from that ancestor to the present. Calculations in parentheses refer to rate estimates for embryonic α and β haemoglobins, respectively.*

Evolutionary period	Age (Myr)	Alpha-HB (NR%)	Beta-HB (NR%)	Myoglobin (NR%)
Vertebrate to gnathostome ancestors	535–420	85.7	67.9	68.8
Gnathostome to amniote ancestors	420–310	54.0(87.7)	91.5	77.8
Amniote to ancestral Neornithes	310–115	14.7(12.8)	14.4	4.0
Amniote to eutherian ancestors	310–85	11.8(10.9)	17.4(14.6)	10.7
Eutherian to anthropoid ancestors	85–40	24.9	41.1	20.3
Anthropoid ancestor to man	40–0	7.0	12.0	18.0
Eutherian ancestor to present	85–0	26.4(17.0)	33.6(20.6)	19.5
Marsupial ancestor to present	85–0	43.2	28.7	9.7
Neornithes ancestor to present	115–0	19.1(10.4)	3.9	15.7
Crocodylid ancestor to present	115–0	13.9	27.8	
Lissamphibian ancestor to present	235–0	39.7		
Anuran ancestor to present	140–0	(30.5)	43.5(31.2)	
Teleost ancestor to present	120–0	41.6		
Selachian ancestor to present	170–0			36.2

towards steeper slow-downs in certain homoiotherms (e.g. hominoids and birds), suggests that Darwinian evolution of proteins occurred. Early in phylogeny most mutations in proteins might have been selectively neutral (Goodman, 1961, 1963, 1985). However, bursts of adaptive substitutions during major radiations of life, shaping dense arrays of lock-and-key interaction sites on proteins, so increased the number of selective constraints that most mutations in proteins are now detrimental (Goodman, 1981, 1985).

Thus, according to this hypothesis, the slow-down in rates resulted from proteins having fewer sites where selectively neutral mutations could accumulate. Take the case of globins: their speeded-up rates in the early vertebrates occurred when selection on duplicated genes shaped separate myoglobin and haemoglobin branches and, just before divergence of Chondrichthyes and Osteichthyes, divided the haemoglobin branch into α and β loci. The fastest-evolving sequence positions in nascent α and β chains were emerging α1β2 contact and Bohr effect sites (Goodman, 1981), the sites which brought about co-operative subunit interactions in heterotetrameric haemoglobin and more efficient delivery of oxygen to respiring tissues. Selection first perfected these important functional sites and then preserved them, as is evident from the fact that later, from the amniote ancestor to the present, they along with haem contact sites were the most slowly evolving, or least variant positions. Similarly, in the jawed vertebrate myoglobin branch, certain functional sites (e.g. salt bridges which stabilize monomeric structure) that were fast evolving in the early vertebrates became very slow-evolving sites from the amniote ancestor to the present (Goodman, 1981).

The β haemoglobin chains of birds evolved at a particularly slow rate, possibly because of an earlier evolution of further functional sites concerned with rapid delivery of oxygen to the flight muscles, which operate at high energetic levels. In contrast, among the later lines of vertebrate descent, we find that globin evolution tended to be fastest in frogs, salamanders, sharks, teleosts and squamates. These globins could conceivably have retained more sites where neutral mutations could occur. Alternatively, vagaries of climate and other fluctuating environmental factors could have favoured almost continuous selection of modifications at less critical functional sites of globins in cold-blooded vertebrates, whereas the maintenance of constant body temperature and other aspects of more efficient homoeostasis buffered the larger-brained, warm-blooded amniotes from external selective pressures.

In the Darwinian view, molecular evolution is part of organismal evolution. During each major radiation of life, natural selection spread novel, beneficial mutations through a stem-lineage. Rates of genomic evolution accelerated in response to new environmental conditions, exerting pressures for new adaptations at organismal and molecular levels. Much later, in its purifying or policing form, natural selection maintained complexes of the beneficial genes at high frequencies throughout the branches descended from the stem-lineage. Accepting this Darwinian view of natural selection as both a conserving and transforming force, the neutralist and selectionist explanations of fast rates of protein evolution are not necessarily in opposition. Relaxation of constraint in a duplicated gene bypasses the force of purifying selection. This opens up the gene for random fixations of mutations but also provides the opportunity for selection of new functions in the protein encoded by that gene. The recurring speed-ups of protein evolution during major radiations of life would then be periods in which, by virtue of gene duplication and Darwinian selection, new or modified proteins evolved; an example is the emergence of histones and members of the calmodulin family in the basal eukaryotes (Goodman, 1981; Baba *et al.*, 1984). Further examples are the myoglobin and α and β haemoglobins that were shaped in the early jawed vertebrates, the two types of α haemoglobins – one expressed in embryonic life and the other in later developmental stages – which emerged in a bony fish ancestor of land vertebrates, and the four to five types of mammalian β haemoglobins expressed in embryonic and later development stages (Goodman *et al.*, 1982*a*, 1984).

Molecular evolution of proteins is not decoupled from organismal evolution. On the grand scale of geological macroevolutionary time, change in molecules and change in gross hard-part anatomy occurred in concert. Gould (1983), in one of his perceptive *Natural History* essays on the fossil record, describes nature's great era of experiments as being about 600 to 300 MyrBP. Life existed then in a multitude of bizarre forms with many basic designs, but after their early efflorescence these forms decreased to today's standardized survivors. The unsurpassed explosion of multicellular life more than 300 MyrBP is reflected in such proteins as haemoglobin and myoglobin which evolved at their fastest rates during that era.

Outlook

With regard to both cladistic pattern and evolutionary process, we conclude that molecular and classical approaches provide broadly overlapping and complementary views of the same large picture of vertebrate phylogeny. Indeed, the molecular tree of vertebrate life revealed so far, sketchy as it is, already validates major findings from the phylogenetic research of palaeontologists and comparative anatomists. The contribution of the new molecular approach to evolutionary history is to extend the frontiers of the discipline and to give a sharper and more accurate picture of the ramifying branches of life. We have only begun to tap the tremendous amount of phylogenetic information stored in DNA. Amino acid sequencing of proteins began in earnest only during the last generation. Now, cloning of genes and nucleotide sequencing of the genic DNA provide almost boundless opportunities to chart for genes and species the exact course of evolutionary descent from common ancestors. We anticipate that as the broader community of systematists and evolutionary biologists comes to take full advantage of the molecular approach, many of our present questions on phylogeny will be solved.

ACKNOWLEDGEMENTS

The research described in this chapter is receiving grant support from the National Science Foundation, National Institutes of Health and the Alfred P. Sloan Foundation. We thank B. F. Koop, M. R. Tennant, P. Janvier and C. Patterson for their invaluable comments and assistance with the research and on the text. We also thank G. Braunitzer, G. Matsuda, T. Kleinschmidt, W. W. de Jong, and J. J. Beintema for providing us with unpublished protein sequences.

REFERENCES

Andres, A.-C., Hosback, H. A. & Weber, R. (1984). Comparative analysis of the cDNA sequences derived from the larval and the adult α_1-globin mRNAs of *Xenopus laevis*. *Biochimica et biophysica Acta*, **781,** 294–301.

Aschauer, H., Sanguansermsi, T. & Braunitzer, G. (1981). Embryonale Hämoglobine des Menschen: die Primärstrukur des ξ-ketten. *Hoppe-Seyler's Zeitschrift für Physiologische Chemie*, **362,** 1159–62.

Aschauer, H., Wiesner, H. & Braunitzer, G. (1984). Zur intrinsischen Sauerstoffaffinität: Die Primärstruktur eines weiteren Ruminantia-Hämoglobins: Methionin in βNA2 eines Stirnwaffenträgers, des Nordland-Elches (*Alces alces alces*). *Hoppe-Seyler's Zeitschrift für Physiologische Chemie*, **365,** 1323–30.

Baba, M. L., Goodman, M., Berger-Cohn, J., Demaille, J. G. & Matsuda, G. (1984). The early adaptive evolution of calmodulin. *Molecular Biology and Evolution*, **1,** 442–55.

Beintema, J. J., Wietzes, P., Weickmann, J. L. & Glitz, D. G. (1984). The amino acid sequence of human pancreatic ribonuclease. *Analytical Biochemistry*, **136,** 48–64.

Bieber, F. A. & Braunitzer, G. (1984). Die embryonalen Hämoglobine vom Hausschwein (*Sus scrofa domestica*). *Hoppe-Seyler's Zeitschrift für Physiologische Chemie*, **365,** 321–34.

Bonde, N. (1975). Origin of 'higher groups': viewpoints of phylogenetic systematics. *Colloques Internationaux du Centre National de la Recherche Scientifique*, **218,** 293–324.

Braunitzer, G., Jelkmann, W., Stangl, A., Schrank, B. & Krombach, C. (1982). Die primäre Struktur des Hämoglobins des indischen Elefanten (*Elephas maximus*, Proboscidea): β2 – Asn. *Hoppe-Seyler's Zeitschrift für Physiologische Chemie*, **363,** 683–91.

Braunitzer, G., Paul, C., Schnek, A. G., Stangl, A. & Schrank, B. (1983*a*). Structure of the llama myoglobin. In *Brussels Hemoglobin Symposium*, ed. A. G. Schnek & C. Paul, pp. 473–8. Bruxelles: Editions de l'Université de Bruxelles.

Braunitzer, G., Schrank, B., Stangl, A. & Wiesner, H. (1980). Respiration at high altitudes, phosphate–protein interaction: sequences of the hemoglobins of the hamster (*Mesocricetus auratus*) and the camel (*Camelus ferus*, Camelidae). *Journal of the Chemical Society of Pakistan*, **2,** 1–7.

Braunitzer, G., Stangl, A., Schrank, B., Krombach, C. & Wiesner, H. (1984). The primary structure of the haemoglobin of the African elephant (*Loxodonta africana*, Proboscidea): asparagine in position 2 of the β-chain. *Hoppe-Seyler's Zeitschrift für Physiologische Chemie*, **365,** 743–9.

Braunitzer, G., Wright, P. G., Stangl, A., Schrank, B. & Krombach, C. (1983*b*). Amino acid sequence of haemoglobin of hippopotamus (*Hippopotamus amphibius*, Artiodactyla). *South African Journal of Science*, **79,** 411–12.

Cracraft, J. (1981). Toward a phylogenetic classification of the Recent birds of the world (class Aves). *Auk*, **98,** 681–714.

Dayhoff, M. O. (1972). *Atlas of Protein Sequence and Structure*, vol. 5. Washington D.C.: National Biomedical Research Foundation.

Dene, H., Goodman, M., Walz, D. A. & Romero-Herrera, A. E. (1983). The phylogenetic position of aardvark (*Orycteropus afer*) as suggested by its myoglobin. *Hoppe-Seyler's Zeitschrift für Physiologische Chemie*, **364,** 1585–95.

Fisher, W. K., Koureas, D. D. & Thompson, E. O. P. (1981). Myoglobins of cartilaginous fishes III. Amino acid sequence of myoglobin of the shark *Galeorhinus australis*. *Australian Journal of Biological Sciences*, **34,** 5–10.

Fitch, W. M. (1970). Distinguishing homologous and analogous proteins. *Systematic Zoology*, **19,** 99–113.

Forey, P. L. (1984). Yet more reflections on agnathan–gnathostome relationships. *Journal of Vertebrate Paleontology*, **4,** 330–43.

Gaffney, E. S. (1984). Historical analysis of theories of chelonian relationship. *Systematic Zoology*, **33,** 283–301.

Gardiner, B. G. (1982). Tetrapod classification. *Zoological Journal of the Linnean Society*, **74,** 207–32.

Gardiner, B. G. (1983). Gnathostome vertebrae and the classification of the Amphibia. *Zoological Journal of the Linnean Society*, **79,** 1–59.

Godovac-Zimmermann, J. & Braunitzer, G. (1983). The amino-acid sequence of northern mallard (*Anas platyrhynchos platyrhynchos*) hemoglobin. *Hoppe-Seyler's Zeitschrift für Physiologische Chemie*, **364,** 665–74.

Godovac-Zimmermann, J. & Braunitzer, G. (1984*a*). Hemoglobin of the adult white stork (*Ciconia ciconia*, Ciconiiformes). *Hoppe-Seyler's Zeitschrift für Physiologische Chemie*, **365,** 1107–13.

Godovac-Zimmermann, J. & Braunitzer, G. (1984*b*). The amino-acid sequence of α^A- and β-chains from the major hemoglobin component of American flamingo (*Phoenicopterus ruber ruber*). *Hoppe-Seyler's Zeitschrift für Physiologische Chemie*, **365,** 437–43.

Goodman, M. (1961). The role of immunochemical differences in the phyletic development of human behavior. *Human Biology*, **33,** 131–62.

Goodman, M. (1963). Man's place in the phylogeny of primates as reflected in serum proteins. In *Classification and Human Evolution*, ed. S. L. Washburn, pp. 204–34. Chicago: Aldine Press.

Goodman, M. (1981). Decoding the pattern of protein evolution. *Progress in Biophysics & Molecular Biology*, **38,** 105–64.

Goodman, M. (1985). Rates of molecular evolution: the hominoid slowdown. *Bioessays*, **3**(1), 9–14.

Goodman, M., Czelusniak, J. & Beeber, J. E. (1985). Phylogeny of primates and other eutherian orders: a cladistic analysis using amino acid and nucleotide sequence data. *Cladistics*, **1,** 171–85.

Goodman, M., Czelusniak, J., Moore, G. W., Romero-Herrera, A. E. & Matsuda, G. (1979). Fitting the gene lineage into its species lineage: a parsimony strategy illustrated by cladograms constructed from globin sequences. *Systematic Zoology*, **28,** 132–63.

Goodman, M., Koop, B. F., Czelusniak, J., Weiss, M. L. & Slightom, J. L. (1984). The η-globin gene: its long evolutionary history in the β-globin gene family of mammals. *Journal of Molecular Biology*, **180,** 802–23.

Goodman, M., Moore, G. W. & Matsuda, G. (1975). Darwinian evolution in the genealogy of haemoglobin. *Nature, London*, **253,** 603–8.

Goodman, M., Romero-Herrera, A. E., Dene, H., Czelusniak, J. & Tashian, R. E. (1982*a*). Amino acid sequence evidence on the phylogeny of primates and other eutherians. In *Macromolecular Sequences in Systematic and Evolutionary Biology*, ed. M. Goodman, pp. 115–91. New York: Plenum Press.

Goodman, M., Weiss, M. L. & Czelusniak, J. (1982*b*). Molecular evolution above the species level: branching pattern, rates, and mechanisms. *Systematic Zoology*, **31,** 376–99.

Gould, S. J. (1983). Nature's great era of experiments. *Natural History*, **92**(7), 12–21.

Gould, S. J. & Eldredge, N. (1977). Punctuated equilibria: the tempo and mode of evolution reconsidered. *Paleobiology*, **3,** 115–51.

Gurnett, A. M., O'Connell, J. & Lehmann, H. (1983). Two rodent myoglobins – viscacha and mole rat. In *Brussels Hemoglobin Symposium*, ed. A. G. Schnek & C. Paul, pp. 467–71. Bruxelles: Editions de l'Université de Bruxelles.

Hardisty, M. W. (1982). Lampreys and hagfishes: analysis of cyclostome relationships. In *The Biology of Lampreys*, vol. 4, ed. M. W. Hardisty & I. C. Potter, pp. 165–260. London: Academic Press.

Heinbokel, N. & Lehmann, H. (1984). The myoglobin of primates: the night monkey, *Aotus trivirgatus* (Cebidae, Platyrrhini, Anthropoidea). *FEBS Letters*, **165,** 46–50.

Janvier, P. (1981). The phylogeny of the Craniata, with particular reference to the significance of fossil 'agnathans'. *Journal of Vertebrate Paleontology*, **1,** 121–59.

Janvier, P. (1984). The relationships of the Osteostraci and Galeaspida. *Journal of Vertebrate Paleontology*, **4,** 344–58.

Janvier, P. (1985). Environmental framework of the diversification of the Osteostraci during the Silurian and Devonian. *Philosophical Transactions of the Royal Society of London, B*, **309,** 259–72.

Jarvik, E. (1980). *Basic Structure and Evolution of Vertebrates*, 2 vols. London: Academic Press.

Jurgens, J. D. (1971). The morphology of the nasal region of Amphibia and its bearing on the phylogeny of the group. *Annale van die Universiteit van Stellenbosch*, **46A,** 1–146.

King, C. R., Shinohara, T. & Piatigorsky, J. (1982). α A-crystallin messenger RNA of the mouse lens: more non-coding than coding sequences. *Science*, **215,** 985–7.

Kleinschmidt, T. & Braunitzer, G. (1982). Die Primärstruktur des Hämoglobins vom ägyptischen Flughund (*Rousettus aegyptiacus*, Chiroptera). *Hoppe-Seyler's Zeitschrift für Physiologische Chemie*, **363,** 1209–15.

Kleinschmidt, T. & Braunitzer, G. (1983*a*). Die Primärstruktur des Hämoglobins vom abessinoschen Kippschliefer (*Procavia habessinica*, Hyracoidea): Insertion von Glutamin in den α-Ketten. *Hoppe-Seyler's Zeitschrift für Physiologische Chemie*, **364,** 1303–13.

Kleinschmidt, T. & Braunitzer, G. (1983*b*). Die Primärstruktur des Hämoglobins vom grossen Tummles (*Tursiops truncatus*, Cetacea). *Biomedica Biochimica Acta*, **42,** 685–95.

Kleinschmidt, T., Nevo, E. & Braunitzer, G. (1984). The primary structure of the hemoglobin of the mole rat (*Spalax ehrenbergi*, Rodentia, chromosome species 60). *Hoppe-Seyler's Zeitschrift für Physiologische Chemie*, **365,** 531–7.

Lalthantluanga, R. & Braunitzer, G. (1981). The primary structure of the β^{I}- and β^{II}-chains of yak hemoglobin (Bovidae). *Hoppe-Seyler's Zeitschrift für Physiologische Chemie*, **362,** 1405–9.

Lalthantluanga, R. & Braunitzer, G. (1982). Complete amino acid sequences of ${}^{I}\alpha$ and ${}^{II}\alpha$ chains of yak hemoglobin. *Indian Journal of Biochemistry and Biophysics*, **19,** 418–20.

Løvtrup, S. (1977). *The Phylogeny of Vertebrata.* London: John Wiley & Sons.

Maeda, N., Zhu, D. & Fitch, W. M. (1984). Amino acid sequences of lower vertebrate parvalbumins and their evolution: parvalbumins of boa, turtle, and salamander. *Molecular Biology and Evolution*, **1,** 473–88.

Maita, T., Hayashida, M. & Matsuda, G. (1984). Primary structures of adult hemoglobins of silvery marmoset, *Callithrix argentatus*, and cotton-headed tamarin, *Saguinus oedipus. Journal of Biochemistry*, **95,** 805–13.

Maruyama, T., Watt, K. W. K. & Riggs, A. (1980). Hemoglobins of the tadpole of the bullfrog, *Rana catesbeiana*. Amino acid sequence of the α chain of a major component. *Journal of Biological Chemistry*, **255,** 3285–93.

Mazur, G. & Braunitzer, G. (1982). Perissodactyla: die Sequenz der Hämoglobine von Wildesel (*Equus hemionus kulan*) und Zebra (*Equus zebra*). *Hoppe-Seyler's Zeitschrift für Physiologische Chemie*, **363,** 59–71.

Mazur, G. & Braunitzer, G. (1984). Perissodactyla: die Primärstruktur der Hämoglobine eines Flachlandtapirs (*Tapirus terrestris*): Glutaminsäure in Position 2 der β-Ketten. *Hoppe-Seyler's Zeitschrift für Physiologische Chemie*, **365,** 1097–106.

Mazur, G., Braunitzer, G. & Wright, P. G. (1982). Die Primärstruktur des Hämoglobins vom Breitmaulnachorn (*Ceratotherium simum*, Perissodactyla): β2 Glu. *Hoppe-Seyler's Zeitschrift für Physiologische Chemie*, **363,** 1077–85.

Meyerhoff, W., Klinger-Mitropoulos, S., Stalder, J., Weber, R. & Knockel, W. (1984). The primary structure of the larval β_1-globin gene of *Xenopus laevis* and its flanking regions. *Nucleic Acids Research*, **12,** 7705–19.

Miyamoto, M. M. & Goodman, M. (1986). Biomolecular systematics of eutherian mammals: phylogenetic patterns and classification. *Systematic Zoology*, **35,** 230–40.

Moore, G. W. (1976). Proof for the maximum parsimony ('red king') algorithm. In *Molecular Anthropology*, ed. M. Goodman & R. E. Tashian, pp. 117–37. New York: Plenum Press.

Moore, G. W., Barnabas, J. & Goodman, M. (1973). A method for constructing maximum parsimony ancestral amino acid sequences on a given network. *Journal of Theoretical Biology*, **38,** 459–85.

Nakamura, S., Takenaka, O. & Takahashi, K. (1983). Fibrinopeptides A and B of baboons (*Papio anubis, Papio hamadryas*, and *Theropithecus gelada*); their amino acid sequences and evolutionary rates and a molecular phylogeny for the baboons. *Journal of Biochemistry*, **94,** 1973–8.

Nelson, G. J. (1969). Gill arches and the phylogeny of fishes, with notes on the classification of vertebrates. *Bulletin of the American Museum of Natural History*, **141,** 475–552.

Niessing, J. (1981). Molecular cloning and nucleotide sequence analysis of adult duck β-globin cDNA. *Biochemistry International*, **2,** 113–20.

Niessing, J. & Erbil, C. (1983). Chromosomal arrangement and the complete nucleotide sequence of the duck α-like globin genes α^A, α^D, and π. In *Brussels Hemoglobin Symposium*, ed. A. G. Schnek & C. Paul, pp. 421–32. Bruxelles: Editions de l'Université de Bruxelles.

Oberthür, W. & Braunitzer, G. (1984). Hämoglobine vom gemeinen Star (*Sturnus vulgaris*, Passeriformes). Die Primärstrukturen der α^A, α^D- und β-Ketten. *Hoppe-Seyler's Zeitschrift für Physiologische Chemie*, **365,** 159–73.

Oberthür, W., Braunitzer, G., Baumann, R. & Wright, P. G. (1983*a*). Die Primärstruktur der α- und β-Ketten der Hauptkomponenten der Hämoglobine des Strausses (*Struthio camelus*) und des Nandus (*Rhea americana*) (Struthioformes). *Hoppe-Seyler's Zeitschrift für Physiologische Chemie*, **364,** 119–34.

Oberthür, W., Braunitzer, G., Grimm, F. & Kosters, J. (1983*b*). Hämoglobine des Steinadlers (*Aquila chrysaetos*, Accipitriformes): die Aminosäure-Sequenz der α^A- und β-Ketten des Hauptkomponente. *Hoppe-Seyler's Zeitschrift für Physiologische Chemie*, **364,** 851–8.

Oberthür, W., Godovac-Zimmermann, J. & Braunitzer, G. (1983*c*). The different evolution of bird hemoglobin chains. In *Brussels Hemoglobin Symposium*, ed. A. G. Schneck & C. Paul, pp. 365–75. Bruxelles: Editions de l'Université de Bruxelles.

Oberthür, W., Wiesner, H. & Braunitzer, G. (1983*d*). Die Primärstruktur der α- und β-Ketten der Hauptkomponente der Hämoglobine der Spaltfussgans (*Anseranas semipalmata*, Anatidae). *Hoppe-Seyler's Zeitschrift für Physiologische Chemie*, **364,** 51–9.

Patient, R. K., Harris, R., Walmsley, M. E. & Williams, J. G. (1983). The complete nucleotide sequence of the major adult β globin gene of *Xenopus laevis*. *Journal of Biological Chemistry*, **258,** 8521–3.

Rodewald, K., Stangl, A. & Braunitzer, G. (1984). Primary structure, biochemical and physiological aspects of hemoglobin from South American lungfish (*Lepidosiren paradoxus*, Dipnoi). *Hoppe-Seyler's Zeitschrift für Physiologische Chemie*, **365,** 639–49.

Romer, A. S. (1966). *Vertebrate Paleontology*, 3rd edn. Chicago: University of Chicago Press.

Romer, A. S. & Parsons, T. S. (1977). *The Vertebrate Body*, 4th edn. Philadelphia: W. B. Saunders & Co.

Rosen, D. E., Forey, P. L., Gardiner, B. G. & Patterson, C. (1981). Lungfishes, tetrapods, paleontology, and plesiomorphy. *Bulletin of the American Museum of Natural History*, **167**, 159–276.

Rücknagel, K. P., Reischl, E. & Braunitzer, G. (1984). Expression von α^{D}-Genen bei Schildkröten, *Chrysemys picta bellii* und *Phrynops hilarii* (Testudines). *Hoppe-Seyler's Zeitschrift für Physiologische Chemie*, **365,** 1163–71.

Schaeffer, B. & Thomson, K. S. (1980). Reflections on agnathan–gnathostome relationships. In *Aspects of Vertebrate History*, ed. L. L. Jacobs, pp. 19–33. Flagstaff: Museum of Northern Arizona Press.

Simpson, G. G. (1945). The principles of classification and a classification of mammals. *Bulletin of the American Museum of Natural History*, **85,** 1–350.

Smith, T. F. & Waterman, M. S. (1981). Identification of common molecular sequences. *Journal of Molecular Biology*, **147,** 195–7.

Söderqvist, T. & Blömback, B. (1971). Fibrinogen structure and evolution. *Naturwissenschaften*, **58,** 16–23.

Stanley, S. M. (1979). *Macroevolution.* San Francisco: Freeman, Cooper & Co.

Stapel, S. O., Leunissen, J. A. M., Versteeg, M., Wattel, J. & de Jong, W. W. (1984). Ratites as oldest offshoot of avian stem – evidence from α-crystallin A sequences. *Nature, London*, **311,** 257–9.

von Wahlert, G. (1968). *Latimeria und die Geschichte der Wirbeltiere: eine evolutionsbiologische Untersuchung.* Stuttgart: Gustav Fischer Verlag.

Watts, D. A., Angelides, T. & Brown, W. D. (1983). The primary structure of myoglobin from Pacific sea turtle (*Chelonia mydas caranigra*). *Biochimica et biophysica Acta*, **742,** 310–17.

Wiley, E. O. (1979*a*). An annotated Linnaean hierarchy, with comments on natural taxa and competing systems. *Systematic Zoology*, **28,** 308–37.

Wiley, E. O. (1979*b*). Ventral gill arch muscles and the interrelationships of gnathostomes, with a new classification of the Vertebrata. *Zoological Journal of the Linnean Society*, **67,** 149–79.

Wiley, E. O. (1981). *Phylogenetics. The Theory and Practice of Phylogenetic Systematics.* New York: John Wiley & Sons.

Yalden, D. W. (1985). Feeding mechanisms as evidence for cyclostome monophyly. *Zoological Journal of the Linnean Society*, **84,** 291–300.

APPENDIX

References for amino acid sequences not cited in Goodman (1981) or Goodman *et al.* (1982*a*, *b*, 1985).

Haemoglobins

Class Mammalia

Alces alces (Aschauer, Wiesner & Braunitzer, 1984)
Balaenoptera acutorostrata (G. Braunitzer, pers. comm.)
Bison bonasus (G. Braunitzer, pers. comm.)
Bos grunniens (Lalthantluanga & Braunitzer, 1981, 1982)
Callithrix argentatus (Maita, Hayashida & Matsuda, 1984)

Ceratotherium simum (Mazur, Braunitzer & Wright, 1982)
Citellus townsendi (G. Braunitzer, pers. comm.)
Elephas maximus (Braunitzer *et al.*, 1982)
Equus hemionus (Mazur & Braunitzer, 1982)
Hippopotamus amphibius (Braunitzer *et al.*, 1983*b*)
Homo sapiens (Aschauer, Sanguansermsi & Braunitzer, 1981)
Hylobates lar (G. Matsuda, pers. comm.)
Loxodonta africana (Braunitzer *et al.*, 1984)
Macaca fascicularis (Dayhoff, 1972)
Mesocricetus auratus (Braunitzer *et al.*, 1980)
Phoca vitulina (G. Matsuda, pers. comm.)
Procavia habessinica (Kleinschmidt & Braunitzer, 1983*a*)
Rousettus aegyptiacus (Kleinschmidt & Braunitzer, 1982)
Spalax ehrenbergi (Kleinschmidt, Nevo & Braunitzer, 1984)
Sus scrofa (Bieber & Braunitzer, 1984)
Tapirus terrestris (Mazur & Braunitzer, 1984)
Tragelaphus strepsiceros (G. Braunitzer, pers. comm.)
Trichechus inunguis (T. Kleinschmidt, pers. comm.)
Tursiops truncatus (Kleinschmidt & Braunitzer, 1983*b*)

Class Aves

Anas platyrhynchos (Godovac-Zimmermann & Braunitzer, 1983)
Anseranas semipalmata (Oberthür, Wiesner & Braunitzer, 1983*d*)
Aquila chrysaetos (Oberthür *et al.*, 1983*b*)
Cairinia moschata (Niessing, 1981; Niessing & Erbil, 1983)
Ciconia ciconia (Godovac-Zimmermann & Braunitzer, 1984*a*)
Phasianus colchicus (Oberthür, Godovac-Zimmermann & Braunitzer, 1983*c*)
Phoenicopterus ruber (Godovac-Zimmermann & Braunitzer, 1984*b*)
Rhea americana (Oberthür *et al.*, 1983*a, c*)
Struthio camelus (Oberthür *et al.*, 1983*c*)
Sturnus vulgaris (Oberthür *et al.*, 1983*c*; Oberthür & Braunitzer, 1984)

Class Amphibia

Rana catesbeiana (Maruyama, Watt & Riggs, 1980)
Xenopus laevis (Patient *et al.*, 1983; Andres, Hosback & Weber, 1984; Meyerhoff *et al.*, 1984)

Class Osteichthyes

Lepidosiren paradoxus (Rodewald, Stangl & Braunitzer, 1984)

Class Reptilia

Chrysemys picta (Rücknagel, Reischl & Braunitzer, 1984)
Phrynops hilarii (Rücknagel *et al.*, 1984)

Myoglobin

Class Mammalia

Aotus trivirgatus (Heinbokel & Lehmann, 1984)
Lagostomus maximus (Gurnett, O'Connell & Lehmann, 1983)
Lama vicugna (Braunitzer *et al.*, 1983*a*)
Orycteropus afer (Dene *et al.*, 1983)
Spalax ehrenbergi (Gurnett *et al.*, 1983)

Class Reptilia

Chelonia mydas (Watts, Angelides & Brown, 1983)

Class Chondrichthyes

Galeorhinus australis (Fisher, Koureas & Thompson, 1981)

Lens α crystallin A

Class Mammalia

Aotus trivirgatus (W. W. de Jong, pers. comm.)
Bassariscus sp. (W. W. de Jong, pers. comm.)
Mus musculus (King, Shinohara & Piatigorsky, 1982)

Class Aves

Anas platyrhynchos (Stapel *et al.*, 1984)
Buteo buteo (Stapel *et al.*, 1984)
Columba livia (Stapel *et al.*, 1984)
Corvus corone (Stapel *et al.*, 1984)
Cygnus olor (Stapel *et al.*, 1984)
Dromaius novaehollandiae (Stapel *et al.*, 1984)
Meleagris gallopavo (Stapel *et al.*, 1984)
Pygoscelis papua (Stapel *et al.*, 1984)
Rhea americana (Stapel *et al.*, 1984)
Struthio camelus (Stapel *et al.*, 1984)

Class Reptilia

Alligator mississippiensis (Stapel *et al.*, 1984)

Fibrinopeptides

Class Mammalia

Alces alces (Söderqvist & Blömback, 1971)
Bos grunniens (R. F. Doolittle, pers. comm.)
Theropithecus gelada (Nakamura, Takenaka & Takahashi, 1983)

Ursus arctos (Söderqvist & Blömback, 1971)

Class Reptilia

Tachydosaurus rugosus (Söderqvist & Blömback, 1971)

Ribonuclease

Class Mammalia

Homo sapiens (Beintema *et al.*, 1984)
Chelydra serpentina (J. J. Beintema, pers. comm.)

7 Macroevolution in the microscopic world

C. R. Woese

Introduction

We understand very little about evolution, particularly the type of evolution involved in the creation of the major taxa, the kingdoms, the phyla and so on. We call this 'macroevolution', to distinguish it from a seemingly different process, 'microevolution', which is characteristic of evolution in the lower taxa. However, the term 'macroevolution' serves more to hide our ignorance than symbolize our understanding.

Macroevolution is one aspect of what is generally called the tempo–mode problem. The entire issue is ensconced in imprecise definition, ignorance and prejudice, and has been the source of endless, sometimes acrimonious, and usually fruitless debate.

What we know today about the tempo–mode problem is little more than we knew 50 years ago (Mayr, 1942; Simpson, 1944; Greenwood, 1979). Evolution sometimes appears to proceed at different rates in different lines of descent, and even at different rates in the same line of descent. Evolution in some cases occurs as a series of small steps over a relatively long time period, while in others the process is a fulminating one, involving drastic changes that occur over a relatively short time span. A rapid evolutionary course seems to spawn unstable lines as well. Some biologists claim there to be a relationship between the tempo of evolution and its mode (Mayr, 1942; Simpson, 1944). In this type of characterization, however, there is no real appreciation, no understanding, of the process being characterized.

The trouble with the tempo–mode problem is that it is generally framed, as it has always been, in terms of fossil (and so morphological) evidence (e.g. the symposium *Tempo and Mode of Evolution from Micropaleontological Data* published in *Paleobiology*, **4,** no. 3 [1984]), which makes the system both ill-defined and intractable. We cannot be sure whether we are dealing with a single class of phenomenon, or a group of superficially similar but basically unrelated phenomena. We cannot manipulate the system experimentally. If we are ever to understand it, the tempo–mode problem will have to be defined in a simpler, more workable system than it now is. This is perhaps the principal reason for asking whether tempo–mode phenomena occur at the level of bacterial evolution. In a microbial system tempo–mode phenomena could undoubtedly be defined in molecular terms, and the system could, at least to some extent, be experimentally manipulated. Recently it has become apparent that bacterial evolution does manifest tempo–mode characteristics. Both its tempo and (to some extent) its mode can be measured in molecular sequence terms, and the two seem to be related in a simple and appealing way (Woese, Stackebrandt & Ludwig, 1985*b*).

Bacterial evolution

Fifteen years ago, nothing to speak of was known about bacterial evolution, except that the bacteria had existed for a very long time, at least 3 billion years, and photosynthetic bacteria were probably among the earliest forms of bacteria (Knoll & Barghoorn, 1978; Lowe, 1980). The bacterial fossil record, though fairly extensive, was not particularly informative; bacterial morphologies are too simple to tell much. Only the lower bacterial taxa could be defined with any assurance, but even some of these, it turns out, were phylogenetic monstrosities. As a result, the science of microbiology developed essentially without an evolutionary framework.

Today, the subject of bacterial evolution has come into its own. Molecular approaches have defined many, if not most, of the major bacterial taxa. At the highest level the bacteria do not constitute a single taxon, a coherent phylogenetic unit. Rather, they fall into two primary kingdoms (or urkingdoms), each as distinct from the other as it is from the eukaryote urkingdom (Woese & Fox, 1977) (Fig. 7.1). The first of these urkingdoms, the eubacteria, is known to contain approximately 10 major groups, each of which should have a taxonomic status equivalent to a eukaryotic phylum or division. (Within a number of these 'phyla', the major subdivisions are also now clearly defined – Woese *et al.*, 1985*c*.) The second urkingdom, the archaebacteria, comprises two primary divisions, each broken into subdivisions, and so on (Fox *et al.*, 1980).

The bacterial urkingdoms appear older than plants and animals combined, a point dramatically seen in sequence terms. The eukaryotic phylogenetic tree defined by cytochrome *c* sequence comparisons is merely one branch of the alpha purple bacteria (Schwartz & Dayhoff, 1978; Woese *et al.*, 1984*a*), which itself is only one of four subdivisions in one of about ten eubacterial 'phyla' (Woese *et al.*, 1984*a, b*, 1985*d*; Oyaizu & Woese, 1985). Moreover, parts of the eukaryotic cell have evolved from bacteria through endosymbiosis; at least its two major organelles have arisen in this way (Margulis, 1970; Stanier, 1970). An understanding of bacterial evolution represents more than just another addition to the body of evolutionary knowledge. Bacterial evolution is both the underpinning for the evolution we now understand and the biological record of the major events in the history of this planet over the last 4 billion years.

Macroevolution on the microscopic level: the mycoplasmas

Since biologists do not agree as to the definition and meaning of macroevolution, it is important to note that in what follows I use the term as used by Simpson, i.e. to denote unexpected drastic and bizarre changes in phenotype, that occur relatively rapidly, etc. (Simpson, 1944).

A bacterial counterpart of macroevolution obviously cannot be defined in terms of the fossil record. However, macroevolution seems to be recognizable in living systems by a disparity between taxonomic classification and phylogenetic position of an organism. Birds and mammals, for example, are the products of macroevolution within the reptiles. Thus, certain reptiles are

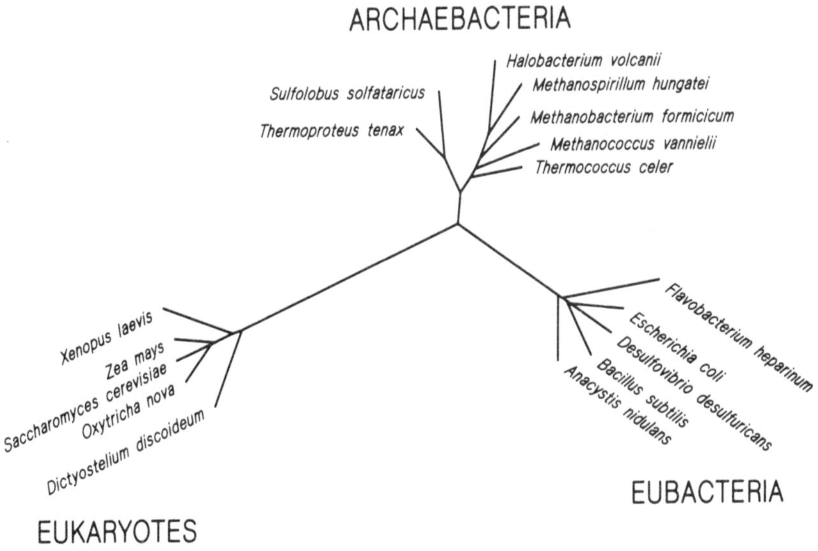

Fig. 7.1 Unrooted universal tree for the three primary kingdoms based on ribosomal RNA sequence comparisons, from Woese & Olsen (1986). The tree was constructed using a standard distance matrix treeing algorithm; distances are in arbitrary units. The eukaryotes include members of all the main groups: animals (clawed toad, *Xenopus*), plants (maize, *Zea*), fungi (yeast, *Saccharomyces*), protists (the ciliate *Oxytricha*) and slime moulds (*Dictyostelium*). The samples of Archaebacteria and Eubacteria are similarly widespread.

phylogenetically related either to birds or to mammals, but are not classified with either of them because their phenotypes have remained reptilian. We shall use this criterion of disparity between taxonomic classification and phylogenetic position to attempt to identify and define macroevolution for the bacteria, and see where it leads us.

One group of bacteria, the mycoplasmas, stands out from all others in terms of the magnitude of the discrepancy between its taxonomic classification and its phylogenetic position. (This discrepancy underlies a long-standing debate regarding the nature of the mycoplasmas.) In phenotype the mycoplasmas are unique. They are unlike all other bacteria in that they have no cell walls (Razin, 1978); they show various biochemical, nutritional and cytological idiosyncrasies; and their genomes tend to be far smaller than typical bacterial genomes. They are classified as *Mollicutes*, which separates them from all other bacteria at a high taxonomic level (Razin, 1978). Explanations of their nature range them from being representatives of a 'pre-bacterial' form of life (that branched from the common line of descent before the common ancestor of all bacteria existed) to simply wall-less states of various standard bacteria (Wallace & Morowitz, 1973).

If the mycoplasmas were not closely related to normal bacteria, then their unusual phenotypic features would merely reflect evolutionary distance from bacteria. However, the mycoplasmas are indeed normal (eu)bacteria phylogenetically: ribosomal RNA sequence comparisons show them to belong to the so-called low G+C (or clostridial) subdivision of the Gram-positive eubacteria, in the subgroup thereof that contains the lactobacilli, streptococci and the genus *Bacillus* (Woese, Maniloff & Zablen, 1980; Woese *et al.*, 1985*c*). In this sense mycoplasmas are nothing more than highly evolved clostridia. They even have several known normal bacterial relatives (Woese *et al.*, 1980, 1985*b*), among them the clostridial species *Clostridium ramosum* and *C. innocuum*, which are closer to them than to other clostridia. However, mycoplasma ribosomal RNA sequences do have at least one unusual characteristic: they are further removed from the rRNAs of outgroup species, such as *Bacillus*, than are the rRNAs of their non-mycoplasma clostridial relatives (Woese *et al.*, 1980). The uniqueness of their phenotype, the discrepancy between their taxonomic and phylogenetic classifications, and the fact that their ribosomal RNA sequences seem to have changed relatively rapidly suggest the mycoplasmas to be the products of some sort of rapid evolution, of macroevolution. We will return to, and further justify, this claim – as well as see other examples of macroevolution on the microscopic level – after familiarizing ourselves with the methods for measuring bacterial genealogies and evolutionary rates.

Molecular chronometers and the measurement of evolutionary rates

The concept of the molecular clock is perhaps the most important addition to evolutionary doctrine since the time of Darwin (Zuckerkandl & Pauling, 1965). It puts the measurement and understanding of evolutionary relationships on a new, more fundamental footing. It demonstrates that there exist two at least semi-independent measures of the evolutionary process, one in terms of phenotypic change (upon which selection can act), the other in terms of the changes in genotype, most of which are in effect selectively neutral (Kimura, 1982, 1983). Genotypic change tends to be 'clock-like' and measures evolution's tempo. Phenotypic change is its mode.

The measurement of bacterial phylogenetic relationships presents problems, because the scope of bacterial evolution is so vast and bacterial morphologies and other characteristics are generally so simple. Approaches perfectly adequate for eukaryotic systems do not work with bacteria. Although microbiologists studied the natural relationships among bacteria for over a century, little progress – and none at the higher taxonomic levels – was recorded until comparative analysis of macromolecules became feasible, particularly sequence comparisons. Even here, however, chronometers effective for eukaryotes, such as cytochrome *c*, did not work well for the bacteria (Woese, 1982).

The ribosomal RNA chronometer

Ribosomal RNA has proven to be a superb device for measuring evolutionary relationships in bacterial (as well as eukaryotic) systems. The reasons are several:

1 Its structure–function constancy. Within an urkingdom, ribosomal RNA sequences are all over 70 per cent homologous (Gupta, Lanter & Woese, 1983; Oyaizu & Woese, 1985; Woese *et al.*, 1985*a*). RNA secondary structure is the same to a first approximation among the three urkingdoms, and the same to a higher approximation within each (Woese *et al.*, 1983). Ribosome function can be reconstituted from ribosomal proteins and rRNAs derived from phylogenetically diverse sources (Nomura *et al.*, 1969). And patterns of change in rRNA sequence tend to be the same from one bacterial group to another (see below).

2 Its capacity to measure both relatively close and very distant evolutionary relationships. Positions in the rRNA sequence exhibit a wide range of rates at which they change. Some have never been observed to vary in composition, a generalization that in a number of cases even includes mitochondrial rRNAs (Woese *et al.*, 1983; Gutell *et al.*, 1985). Other positions vary in composition between species within the same genus (Woese *et al.*, 1983). The difference in rate at which these more conserved vs more variable positions change is, by conservative estimate, over 100-fold.

3 Its relatively large size (1500+ nucleotides). This makes rRNA a 'smoother', more reliable chronometer than smaller molecules such as cytochrome *c* and 5S rRNA (Woese, 1982).

The ribosomal RNA molecule seems to have two important 'chronometric properties'. The first is that the observed changes in sequence are by and large *selectively neutral*. This is inferred from the above discussed constancy of structure–function. (Selective neutrality does not mean that *individual* base changes are necessarily neutral, however. It is the combinations of them that occur naturally that are taken to be so. For example, the replacement of an A–U pair in a helix by a C–G pair could be a selectively neutral change, but the isolated component changes, i.e., A → C or U → G, are not likely to be.) By definition, selectively neutral changes are independent of the evolutionary course in a line of descent, merely reflecting the mutational state of cell, i.e. the probability of fixing a neutral mutation equals the mutation rate in the line of descent (Kimura, 1982, 1983).

The second characteristic is that many of the changes occurring in ribosomal RNA sequence appear to be *higher-order* functions of the mutation rate. This is inferred from the more than 100-fold difference in the rate at which individual base replacements occur in ribosomal RNA (see above), plus the fact that this rate correlates with structural features in the ribosomal RNA (Gupta *et al.*, 1983; Woese *et al.*, 1983). The base pair replacement cited above (A–U → C–G) is one example of a general class of secondary and tertiary structural elements in rRNA, changes which could require simultaneous replacement of two or more component nucleotides (if the overall change is to remain selectively neutral). The occurrence of such changes *relative* to first-order changes varies with the mutation rate in a line of descent; for sufficiently low rates, the higher-order changes would occur with a negligible frequency *relative to* the first-order changes; as the rate increases, the higher-order changes become appreciable by the same metric.

These two characteristics make the ribosomal RNA a powerful chronometer, for, speaking loosely, they allow the amount of change (sequence distance) and the rate of change to be measured independently. This, as we shall see, is tantamount to measuring both the tempo and the mode of evolution.

Table 7.1. *Families of highly conserved oligonucleotides in 16S ribosomal RNA*

The individual families and their approximate positions in the 16S rRNA sequence are listed in the first (sequence) column. The remaining columns show the percentage occurrence of the various oligonucleotides in a given family in each of the major eubacterial taxa. Abbreviations for the various eubacterial taxa are as follows: Gm+ – the Gram-positive eubacteria of (L) low G+C content DNA or (H) high G+C content DNA (Fox *et al.*, 1980); PUR – the alpha (A), beta (B) and gamma (C) subdivisions of the purple bacteria (Woese *et al.*, 1984*a*, 1984*b*, 1985*d*); DSM – delta subdivision of the purple bacteria; sulphur- and sulphate-reducing bacteria (S), myxobacteria (M) and bdellovibrios (B) (Ludwig *et al.*, 1983; Hespell *et al.*, 1984; Fowler *et al.*, in preparation); SPR – spirochaetes and relatives (Paster *et al.*, 1984); S – spirochaetes and treponemes, L – leptospiras, H – anaerobic halophiles (Oren *et al.*, 1984*a*, *b*; Paster *et al.*, 1984); CFB – flavobacteria and relatives (F) and bacteroides (B) (Paster *et al.*, 1985); CY – cyanobacteria (Bonen & Doolittle, 1976, 1978); GN – green sulphur bacteria (Gibson, Stackebrandt & Woese, 1985); CX – green non-sulphur bacteria and relatives (Gibson *et al.*, 1985); RD – radiation-resistant micrococci (Brooks *et al.*, 1980); and PL – planctomyces (Stackebrandt *et al.*, 1984). The number of species catalogued in each group is listed under the corresponding column heading. Bases in brackets are alternatives to one another. Y = pyrimidine, R = purine. Every oligonucleotide is preceded by an implied G residue, except where preceded by dots. Dots in percentage columns mean no occurrences.

	Percentage occurrence of oligonucleotide																	
	Gm+		PUR			DSM			SPR			CFB		CY	GN	CX	RD	PL
	L	H	A	B	C	S	M	B	S	L	H	B	F					
Sequence	(94)	(61)	(23)	(24)	(51)	(10)	(5)	(3)	(16)	(2)	(4)	(10)	(12)	(9)	(4)	(4)	(3)	(8)
170																		
CUAAUACCG	74	15	57	83	79	70	40	67	6	100	.	.	.	44	.	.	.	.
UCUAAUACYG	.	70	.	.	.	.	.	.	.	.	.	.	.	.	.	.	.	.
AYUAAUACCG	3	2	13	17	17	.	.	.	6	.	.	30	8	.	.	.	.	13

Table 7.1. (*contd*)

Sequence	Gm+ L (94)	Gm+ H (61)	PUR A (23)	PUR B (24)	PUR C (51)	DSM S (10)	DSM M (5)	DSM B (3)	SPR S (16)	SPR L (2)	SPR H (4)	CFB B (10)	CFB F (12)	CY (9)	GN (4)	CX (4)	RD (3)	PL (8)
AYUAAYACYY	1	.	.	.	.	.	.	.	.	.	.	40	75	.	.	.	.	.
UAAUACCG	1	.	13	.	.	.	20	.	56	.	.	.	.	.	.	.	.	.
CUAAUACR	1	5	.	.	.	.	20	.	.	.	.	.	.	.	.	.	100	.
315																		
YCACAYYG	98	95	96	92	100	100	100	67	50	50	100	.	.	100	100	.	.	.
RCACAYUG	.	2	.	.	.	.	.	33	50	.	.	.	.	.	.	.	.	13
UCCCCCACAUUG	.	.	.	.	.	.	.	.	.	.	.	100	.	.	.	.	.	.
ACCCCCCACACUG	.	.	.	.	.	.	.	.	.	.	.	.	58	.	.	.	.	.
. [AU] CUCACUG	.	.	.	.	.	.	.	.	.	.	.	.	.	.	.	25	.	25
-YCACAG	.	.	.	.	.	.	.	.	.	.	.	.	.	.	.	.	100	.
365																		
AAUCUUC	55	.	4	.	.	.	.	.	50	.	100	.	.	.	.	25	100	63
AAUCUUR	4	.	34	4	.	10	20	.	.	100	.	.	.	.	.	.	.	.
AAUCUUU	.	.	.	.	.	.	.	.	.	.	.	.	.	.	.	.	.	37
AAUAUUC	.	.	.	.	.	.	.	.	50	.	.	.	.	.	.	.	.	.
AAUAUUG	26	93	53	.	90	30	20	100	.	.	.	100	100	.	100	.	.	.
AAUUUUC	2	.	.	.	.	.	.	.	.	.	.	.	.	100	.	75	.	.
AAUUUUG	.	.	.	96	.	60	60	.	.	.	.	.	.	.	.	.	.	.
510																		
CUAACUYYG	.	.	91	.	83	90	100	67	.	.	.	60	100	89	100	50	100	.
CUAAUUCCG	.	.	.	.	.	.	.	.	.	.	.	40	.	11	.	.	.	.
CYAACUACG	98	100	9	100	12	10	.	.	19	100	50	.	.	.	.	50	.	.
CYAAUUACG	.	.	.	.	.	.	.	.	81	.	50	.	.	.	.	.	.	.

Column headings are grouped under "Percentage occurrence of oligonucleotide".

Table 7.1. (*contd*)

Sequence	Gm+ L (94)	Gm+ H (61)	PUR A (23)	PUR B (24)	PUR C (51)	DSM S (10)	DSM M (5)	DSM B (3)	SPR S (16)	SPR L (2)	SPR H (4)	CFB B (10)	CFB F (12)	CY (9)	GN (4)	CX (4)	RD (3)	PL (8)
	Percentage occurrence of oligonucleotide																	
535																		
UAAUACG	95	89	100	100	87	90	20	67	12	100	75	100	100	89	.	25	100	13
UAAUACAG	.	.	.	.	13	.	60	.	.	.	.	.	.	.	.	25	.	.
795																		
AUACCCUG	99	98	100	100	98	90	100	100	100	50	.	100	100	.	100	50	.	.
AUACCCCG	1	.	.	.	2	10	.	.	.	50	100	.	.	.	.	.	.	100
AUACCCCUG	.	.	.	.	.	.	.	.	.	.	.	.	.	67	.	.	.	.
AUACCCG	.	.	.	.	.	.	.	.	.	.	.	.	.	.	.	50	100	.
815																		
UAAACG	99	98	91	.	89	100	100	100	94	.	100	100	58	100	100	100	.	88
CCCUAAACG	.	.	.	100	11	.	.	.	.	100	.	.	33	.	.	.	.	.
UAAACUAUG	.	.	4	.	.	.	.	.	.	.	.	.	.	.	.	.	.	.
UAAACAAUG	.	.	.	.	.	.	.	.	.	.	.	.	.	.	.	.	.	12
CCCUAAACUAUG	.	.	.	.	.	.	.	.	.	.	.	.	8	.	.	.	.	.
. CACCCUAAACG	.	.	.	.	.	.	.	.	.	.	.	.	.	.	.	.	100	.
910																		
AAACUCAAAG	92	.	4	32	8	20	.	.	100	100	100	100	100	100	100	.	100	100
AUUAAAACUCAAAG	6	.	74	55	13	20	.	33	.	.	.	.	.	.	.	.	.	.
ACUAAAACUCAAAG	.	.	.	.	2	30	100	67	.	.	.	.	.	.	.	.	.	.
UUAAAACUCAAAG	1	.	22	14	12	.	.	.	.	.	.	.	.	.	.	.	.	.
CUAAAACUCAAAG	.	100	.	.	6	30	.	.	.	.	.	.	.	.	.	25	.	.
CUUAAAACUCAAAG	.	.	.	.	.	.	.	.	.	.	.	.	.	.	.	75	.	.
. . AAACUCAAAUG	.	.	.	.	60	.	.	.	.	.	.	.	.	.	.	.	.	.

Table 7.1. (*contd*)

Sequence	Gm+ L (94)	Gm+ H (61)	PUR A (23)	PUR B (24)	PUR C (51)	DSM S (10)	DSM M (5)	DSM B (3)	SPR S (16)	SPR L (2)	SPR H (4)	CFB B (10)	CFB F (12)	CY (9)	GN (4)	CX (4)	RD (3)	PL (8)
960																		
UUUAAUUCG	98	2	100	36	100	100	>40	100	100	100	100	100	100	100	100	100	100	.
AUUAAUUCG	.	74	.	64	.	.	.	.	.	.	.	.	.	.	.	.	.	.
CUUAAUUCG	1	25	.	.	.	.	.	.	.	.	.	.	.	.	.	.	.	100
985																		
. AACCUUACCAR	87	66	87	.	19	10	.	.	38	.	75	.	42	100	.	25	100	.
. AACCUUACCYR	12	34	4	100	71	90	100	100	63	100	.	100	50	.	100	50	.	.
. AACCUUACCG	1	.	.	.	6	.	.	.	.	.	25	.	8	.	.	.	.	.
. AACCUUACCAY	.	.	8	.	4	.	.	.	.	.	.	.	.	.	.	.	.	.
. AACCUUAUCCY	.	.	.	.	.	.	.	.	.	.	.	.	.	.	.	.	.	100
1110																		
CAACCYYCR	1	54	65	.	4	10	60	33	.	.	.	.	8	89	.	25	.	.
CAACCYYUR	94	46	4	100	90	90	40	67	100	100	75	60	92	.	100	50	100	38
CAACCYACR	.	.	31	.	.	.	.	.	.	.	.	.	.	11	.	.	.	.
CAACCYYUY	.	.	.	.	2	.	.	.	.	.	.	40	.	.	.	.	.	.
AAACCCYUR	.	.	.	.	.	.	.	.	.	.	.	.	.	.	.	.	.	62
1200																		
. . UCCUCAUG	.	.	96	96	.	50	100	100	.	100	25	.	.	.	100	.	.	75
. . UCAUCAUG	98	90	4	4	100	40	.	.	100	.	75	.	50	33	.	.	.	25
UCAAAUCAG	1	.	.	.	.	.	.	.	.	.	.	100	.	.	.	.	.	.
. . UCAUCACG	.	.	.	.	.	.	.	.	.	.	.	.	50	.	.	.	.	.

Table 7.1. (*contd*)

Sequence	Gm+ L (94)	Gm+ H (61)	PUR A (23)	PUR B (24)	PUR C (51)	DSM S (10)	DSM M (5)	DSM B (3)	SPR S (16)	SPR L (2)	SPR H (4)	CFB B (10)	CFB F (12)	CY (9)	GN (4)	CX (4)	RD (3)	PL (8)
1210																		
CCCYUUAU	98	95	.	.	.	.	.	.	.	.	100	.	.	.	.	.	.	.
CCCCUUAC	.	5	.	.	.	.	.	.	.	.	.	.	.	89	.	.	.	.
CCCUAUAU	2	.	.	.	.	.	.	.	.	.	.	.	.	.	.	.	.	.
CCCUUAU	.	.	13	100	23	50	40	100	94	.	.	.	.	.	.	.	.	.
CCCUUAC	.	.	87	.	75	20	.	.	6	.	.	100	75	.	100	50	.	.
CCUUUAU	.	.	.	.	2	30	60	.	.	100	.	.	25	.	.	.	.	100
UCCUUAC	.	.	.	.	.	.	.	.	.	.	.	.	.	.	.	.	100	.
1225																		
CYACACACG	98	.	100	35	100	100	100	100	100	50	100	90	91	89	50	.	100	.
CAACACACG	.	.	.	.	.	.	.	.	.	50	.	.	.	.	50	.	.	.
ACACACG	.	.	.	.	.	.	.	.	.	.	.	10	.	11	.	.	.	.
ACACACACG	.	.	.	.	.	.	.	.	.	.	.	.	.	.	.	100	.	.
CUUCACACG	.	.	.	65	.	.	.	.	.	.	.	.	.	.	.	.	.	.
CUUCACACAUG	.	16	.	.	.	.	.	.	.	.	.	.	.	.	.	.	.	.
CUUCACG	.	56	.	.	.	.	.	.	.	.	.	.	.	.	.	.	.	.

Percentage occurrence of oligonucleotide

Evolutionary characteristics of bacterial ribosomal RNAs

In normal bacterial lines

A partial sequencing method (oligonucleotide cataloguing) has been used over the past 15 years to characterize the small subunit rRNAs of more than 400 bacterial species, providing an overview of bacterial phylogeny (Fox *et al.*, 1980). These data also give a feeling for the tempo and what appears to be the mode of bacterial evolution as manifest in ribosomal RNA sequences. (The nature of the data, i.e. small sequence fragments, confines one for the most part to the more highly conserved areas of the sequence, however, where families of related oligonucleotides that cover most eubacterial or archaebacterial phyla can be recognized.)

For any of the conserved locales in the 16S rRNA sequence (that are covered by oligonucleotides of appreciable size), a remarkably small family of oligonucleotides will account for most if not all variation encountered. Table 7.1 shows examples of these oligonucleotide families that cover the areas of high sequence conservation in eubacteria (Woese *et al.*, 1985*c*). Note the small number of oligonucleotide sequences needed in each family to encompass the vast majority of catalogues; also note that most of the variants in each family have arisen more than once. The extent to which this last statement is true is not fully apparent from the data as presented. For example, the detailed phylogeny for the gamma purple bacteria demands that the version AUUAAUACCG at position 170 has arisen three times within that group (Woese *et al.*, 1985*d*); it has arisen twice within the beta purple bacteria by the same argument (Woese *et al.*, 1984*b*). AACCUUACCUR (position 985) appears to have arisen (from AACCUUACCAR) at least five times within the Gram-positive group (unpublished analysis), and two or three times in the gamma purple bacteria.

Table 7.2 (left-hand side) provides a narrower but more detailed look at the families of conserved oligonucleotides in normal groups. The left-hand side of the table shows the distribution of oligonucleotides in families of conserved sequences for five groups of typical eubacteria, the alpha, gamma and delta (i.e. sulphate reducers and relatives) subdivisions of the purple bacteria (Woese *et al.*, 1984*a*, 1985*c*, *d*; Oyaizu & Woese, 1985), and (in most detail) the genera *Bacillus* and *Lactobacillus*, the last two being emphasized because of their relatively close relationship to the mycoplasma group (Woese *et al.*, 1980, 1985*b*).

For each family in Table 7.2 the listed oligonucleotides cover the vast majority of eubacterial 16S rRNA catalogues – generally well over 90 per cent

Table 7.2. *Families of highly conserved sequences examined in detail for several groups of organism*

Format similar to Table 7.1. The approximate position of a family in 16S rRNA sequence and the oligonucleotides in it are listed in columns 1 (pos) and 8 (Sequence) respectively. The various groups of organisms are listed as follows: the delta (sulphur reducing eubacteria), gamma and alpha subdivisions of the purple bacteria (columns 2, 3 and 4); the genera *Bacillus* (column 5) and *Lactobacillus* (column 6), the mycoplasmas (column 10) and the mitochondria (column 11). The number of species in each group is given below the corresponding column heading. For columns 2, 3 and 4, occurrence is given as number of catalogues, not percentage, except for the two families (pos 525, 1520) in which some catalogues were not screened for the sequences in the family; in these cases percentage is used. For the remaining groups (columns 5, 6, 10, 11), occurrence is listed individually, one subcolumn for each species in the group. In the mycoplasma group abbreviations for the species are as follows: g– *Mycoplasma gallisepticum*, s– *Spiroplasma citri*, c– *M. capricolum*, a– *Acholeplasma laidlawii*, i– *Clostridium innocuum*, and r– *C. ramosum* (Woese *et al.*, 1980). The mitochondria in order of occurrence are as follows: p– plant (Spencer, Schnare & Gray, 1984), c– ciliate (Seilhammer, Olsen & Cummings, 1984), f– fungal (Sor & Fukuhara, 1980), and a– animal (Eperon, Anderson & Nierlich, 1980). The 'avg' column gives the per cent occurrence for the oligonucleotide in question for all eubacterial catalogues (about 350); the 'ori' column gives the estimated number of phylogenetically independent occurrences of the sequence in question – estimated from the phylogenetic tree for the eubacteria (unpublished analysis) – 'N' in this column means that the sequence is the dominant one in eubacteria, and is inferred to be ancestral. In columns 5, 6, 10 & 11, a '+' signifies a high-occurrence or ancestral variant, 'x' indicates a minor occurrence variant (four or fewer known independent occurrences), and 'u' indicates a unique variant (among eubacteria). A '0' is inserted (at the first entry in a family) when a given catalogue or sequence does not contain any of the listed members of a family, for whatever reason. Positions of known post-transcriptionally modified nucleosides are indicated by lower-case letters. Other abbreviations and conventions are as used in Table 7.1.

1 pos	2 sulph (10)	3 gamma (52)	4 alpha (23)	5 *Bacillus* (20)	6 *Lactobacillus* (12)	7 avg	8 Sequence	9 ori	10 Mycoplasma (6) g s c a i r	11 mito (4) p c f a
20	8+	21+	23+	++++++ ++++++++ ++++++	++++++ +.++++	87	AUCCUG	N	++++ ++	+000
	2x	.	.	....	....	2	AUUCUG	4	.... ..	....
		30u	.	....	....	9	AUCAUG	1	.... ..	....
		.	.	....	...... .u....	<1	AUCUUG	1	.... ..	....
50	10+	52+	23+	....	....	62	CYUAACACAUG	N	.... ..	+000
	.	.	.	++++++ ++++++++ ++++++	++++++ ++++++	21	CYUAAUACAUG	6	++++ ++	....
		.	.	....	....	7	CYUUACACAUG	1	.... ..	....
270	9+	46+	20+	++++++ ++++++++ ++++++	+.++++ ++++++	89	CYYACCAAG	N	+.++ ++	0000
	.	6+	.	....	....	4	CYYACCUAG	6	.... ..	....
	.	.	1+	....	.+....	3	CYYACCAUG	7	.+.. ..	....
	.	.	1x	....	....	3	CACCAAG	2	.... ..	....
	1x	.	1x	....	....	1	CYYACUAAG	2	.... ..	....
315	9+	52+	22+	++++++ ++++++++ ++++++	+++++. ++++++	85	YCACAYUG	N	.+.+ ++	+000
	1+	.	.	....	....	2	YCACAYCG	4	..+. ..	....
	.	.	.	....	....	2	RCACAYUG	5	.... ..	....
	.	.	.	....	.....x	<1	YCACAAUG	2	.... ..	....
	.	.	.	....	....	<1	AAUAACCACAAUG	1	u... ..	....
	.	.	1u	....	....	<1	CAACAYUG	1	.... ..	....
340	10+	52+	17+	++++++ ++++++.. ++++++	....+....	82	ACUCCUACG	N	.+.. .+	0000
	.	.	6+	....++	+++++. +.++++	13	.AAACUCCUACG	>8	...+ +.	....
	.	.	.	....	....	<1	AACUCCUACG	2	..x. ..	....
	.	.	.	....	.....x	1	.AUACUCCUACG	3	x... ..	....
510	1+	5+	2+	++++++ ++++++++ ++++++	+++.++ ++++++	58	CYAACUACG	N	.... ++	.000
	9+	44+	21+	....	....	35	CUAACUYYG	>5	.... ..	+...
	.	.	.	....	...x..	1	CUAAAUACG	3	.... ..	....
	.	.	.	....	....	1	CUAACUAUG	1	.uuu ..	....
	.	.	.	....	....	<1	ACUAACUAUG	1	u... ..	....

Table 7.2. (*contd*)

1 pos	2 sulph (10)	3 gamma (52)	4 alpha (23)	5 *Bacillus* (20)	6 *Lactobacillus* (12)	7 avg	8 Sequence	9 ori	10 Mycoplasma (6) g s c a i r	11 mito (4) p c f a
525	100%	100%	23+	++++++ ++++++++ ++++++	++++++ ++++++	99	CCgCG	N	.+++ ++	++.0
	.	.	.	....	....	<1	UCgCG	1	u... ..	..x.
535	9+	45+	23+	++++++ ++++++++ +++0++	++++++ ++++++	88	UAAUACG	N	.... ++	0000
	.	7x	.	....	....	3	UAAUACAG	3	.... ..	....
	.	.	.	....	....	1	UAAUACAUAG	1	uuuu ..	....
785	8+	49+	23+	++++++ ++++++++ ++++++	++++++ ++++++	>80	CRAACAG	N	...+ ..	+000
	.	.	.	....	....	2	CAAAUAG	1	uuu. uu	....
795	9+	51+	23+	++++++ ++++++++ ++++++	++++++ ++++++	90	AUACCCUG	N	...+ ..	+0.0
	1+	1+	.	....	....	3	AUACCCCG	6	.... ..	....
	.	.	.	....	....	1	AUACCCG	2	.... ..	..x.
	.	.	.	....	....	2	AUACCCUAG	1	uuu. uu	....
805	8+	49+	21+	++++.+ ++++.++. ++++++	..0.+.	70	UCCACG	N	.+.+ ++	.000
	.	2+	.	....+. ....+..+	+....+ ++++++	11	UCCAUG	>5	.... ..	+...
	.	.	1x	....	....	1	UCUACG	4	..x. ..	....
	.	.	.	....	...x..	1	UCCACACYG	3	x... ..	....
	.	.	.	....	.u....	<1	UCCAUACYG	1	.... ..	....
815	10+	46+	21+	++++++ ++++++++ ++++++	++++++ ++++++	87	UAAACG	N	++++ ++	+..0
	.	6+	.	....	....	11	CCCUAAACG	7	.... ..	....
	.	.	1u	....	....	<1	UAAACUAUG	1	.... ..	..x.
	.	.	1u	....	....	<1	UAAAAG	1	.... ..	.x..
910	2+	4+	1+	++++++ ++++++++ ++++++	++++++ +.++++	46	AAACUCAAAG	N	.+++ ++	+00.
	8+	17+	22+	....	....	42	RYUAAAACUCAAAG	6	.... ..	....
	.	31u	.	....	....	9	..AAACUCAAAUG	1	.... ..	....
	.	.	.	....	....	<1	AAACUCAAACG	1	u... ..	....
	.	.	.	....	...... .u....	<1	AAACUCAAAAG	1	.... ..	....
	.	.	.	....	....	1	CUUAAAACUCAAAG	1	.... ..	...x

Table 7.2. (*contd*)

1 pos	2 sulph (10)	3 gamma (52)	4 alpha (23)	5 *Bacillus* (20)			6 *Lactobacillus* (12)		7 avg	8 Sequence	9 ori	10 Mycoplasma (6)		11 mito (4)
												g s c a	i r	p c f a
935	10+	52+	23+	++++++	++++++++	++++++	++++++	++++++	97	CACAAG	N	++++	++	+000
960	10+	52+	23+	++++++	++++++++	++++++	++++++	++++++	77	UUUAAUUCG	N	.+++	++	+0+0
	.	.	.	......	....	......	......	......	18	AUUAAUUCG	2	....	..	....
	.	.	.	......	....	......	......	......	5	CUUAAUUCG	3	x...	..	....
985	.	10+	20+	++++++	++++++++	++++++	++++++	++++++	55	AACCUUACCAR	N	.++.	++	..00
	10+	37+	1+	......	....	......	......	......	32	AACCUUACCYR	>8	....	..	....
	.	.	.	......	....	......	......	......	8	AAAAACCUUACCYR	3	x...	..	....
	.	3x	.	......	....	......	......	......	2	AACCUUACCG	4	....	..	....
	.	2x	2x	......	....	......	......	......	1	AACCUUACCAY	2	....	..	....
	.	.	.	......	....	......	......	......	<1	NAAAACCUUACCAR	1	...u	..	uu..
1110	9+	47+	1+	++++++	++++++++	++.+++	++++++	++++++	73	CAACCYYUR	N	++++	++	.+00
	1+	2+	15+	......	....	..+...	......	......	21	CAACCYYCR	>6	....	..	....
	.	.	7x	......	....	......	......	......	2	CAACCYACR	2	....	..	....
	.	.	.	......	....	......	......	......	2	AAACCCYYR	1	....	..	u...
1200	4+	.	1+	++++++	++++++++	++++++	++++++	+++++0	70	..UCAUCAUG	N	++.+	++	0000
	5+	52+	22+	......	....	......	......	......	19	..UCCUCAUG	>5	....	..	....
	.	.	.	......	....	......	......	......	3	UCAAAUCAG	2	..x.	..	....
1210	10+	52+	23+	......	....	...0..	......	......	49	CCYUUAYR	N	....	..	+000
	.	.	.	++++++	++++++++	+++.++	++++++	++++++	50	CCCYUUAYR	>2	++++	++	....
1240	6+	49+	23+	++++++	++++++++	..++++	++++++	++++++	82	CUACAAUG	N	+++.	.+	+.00
	3+	3+	.	......	....	++....	......	......	6	UACUACAAUG	6	....	+.	....
	.	.	.	......	....	......	......	......	3	UAAUACAAUG	3	....	..	....
	.	.	.	......	....	......	......	......	1	AUACAAUG	2	...x	..	....
	.	.	.	......	....	......	......	......	3	UUACAAUG	1	....	..	.x..
1320	10+	52+	22+	++++++	++++++.+	++++++	++++++	++++++	87	CAACUCG	N	.+++	++	+000
	.	.	.	......	......+.	......	......	......	5	CAACCCG	6	....	..	....
	.	.	.	......	....	......	......	......	4	AAACUCG	>3	....	..	....
	.	.	.	......	....	......	......	......	<1	CAAUUCG	2	x...	..	....

Table 7.2. (*contd*)

1 pos	2 sulph (10)	3 gamma (52)	4 alpha (23)	5 *Bacillus* (20)	6 *Lactobacillus* (12)	7 avg	8 Sequence	9 ori	10 Mycoplasma (6) g s c a i r	11 mito (4) p c f a
1350	8+	51+	23+	++++++ ++++++++ ++++++	++++++ ++++++	98	UAAUCG	N	++++ ++	+++0
	2u	.	.	....	....	1	UAAUUCG	1	.... ..	....
1375	9+	51+	22+	++++++ +++++++0 ++++++	++++++ ++++++	86	AAUACG	N	++++ ++	0000
1380	10+	52+	23+	++++++ ++++++++ ++++++	++++++ ++++++	94	UUCCCG	N	.... ..	0000
	.	.	.	....	....	3	UUCUCG	3	xxxx xx	....
1400	10+	52+	23+	++++++ ++++++++ ++++++	++++++ ++++++	>99	UACACACCG	N	++++ ++	+000
1500	8+	52+	23+	++++++ ++++++++ ++++++	++++++ ++++++	>99	UAACAAG	N	++++ ++	+00+
1520	100%	91%	23+	....	++++++ ++++++	34	a aCCUG	N	.... ..	+0+.
	.	9%	.	++++++ ++++++++ ++++++	....	62	a aG	>8	.... ..	....
	.	.	.	....	....	2	a aCG	2	xxxx x.	....
	.	.	.	....	....	1	a aAg	2	.... ..	...x
	.	.	.	....	....	<1	aACG	–	.... .u	....

(see 'avg' column). The estimated number of phylogenetically independent occurrences of each sequence is given in the 'ori' column – on the basis of which the occurrences of the oligonucleotides are designated '+' (ancestral or relatively high occurrence variant), 'x' (phylogenetically rare variant) or 'u' (phylogenetically unique variant) (see Table 7.2 caption). For normal bacteria, the variant in each family is almost always of the ancestral or high-occurrence (+) type. In the group of gamma purple bacteria, for example, only 6 per cent of the occurrences involve minor- or unique-occurrence oligonucleotides (in which latter category are also scored those cases where a representative oligonucleotide has not been found for whatever reason; in these cases a 0 is placed in the corresponding column for the first entry in the family). For the alpha subdivision this number is 3 per cent; it is 6 per cent for the sulphate reducers and relatives (delta subdivision). The average catalogue, then, possesses one of the main variants in over 90 per cent of the oligonucleotide families.

The two normal groups examined in most detail, *Bacillus* and *Lactobacillus*, show the same trend. Of the 20 *Bacillus* species catalogued, 18 show a high occurrence (+) variant in *every* family, as do 6 of the 12 *Lactobacillus* catalogues also. Each of the exceptions (2 *Bacillus* and 6 *Lactobacillus* catalogues) shows minor or unique variants (or is missing a member) in only one or two of the families (see Table 7.2). Given the relatively large numbers of catalogues involved, it is safe to conclude that, for normal lines of bacterial descent, the amount and range of variation in these families of highly conserved oligonucleotides is extremely limited.

In rapidly evolving lines

By (ribosomal RNA) sequence distance measure, the mycoplasmas are rapidly evolving lines of descent. Table 7.3 is a homology matrix for a collection of eubacterial 16S rRNA sequences that includes two mycoplasmas and one of their close non-mycoplasma relatives, *Clostridium innocuum*. Analysis of these data shows the mycoplasmas to group with one another and with *C. innocuum*, while *Bacillus subtilis* is a slightly more distant relative (see Fig. 7.2). However, the mycoplasmas tend to be further removed from outgroup species than is *B. subtilis*. The relationships are more striking when expressed in terms of the binary association coefficient S_{AB} (upper right-hand triangle of Table 7.3 – Fox, Peckman & Woese, 1977). [S_{AB} is the sum of bases in all oligonucleotides (hexamer and larger) *common* to any pair of oligonucleotide catalogues, divided by half the sum of all bases in *all* oligonucleotides (hexamer and larger) in both catalogues (Fox *et al.*, 1977). Because the numerator of this expression is to a large extent determined by stretches of bases from highly conserved areas in the rRNA sequence, the S_{AB} value is more sensitive to changes in highly conserved areas than is overall sequence homology.]

By the same measure (rRNA sequence distance), the mitochondria, particularly the animal, ciliate and fungal versions, are indeed spectacular. These three show *far* greater distance from outgroup species than do their normal relatives (Yang *et al.*, 1985). [The closest eubacterial relatives of the mitochondria are the so-called alpha purple bacteria (Yang *et al.*, 1985), a group represented in Table 7.3 by *Agrobacterium tumefaciens*.]

The rapidly evolving lines exhibit a very different pattern in the families of conserved sequences than that just seen for normal bacteria. The right-hand side of Table 7.2 presents data for four mycoplasmas, two of their clostridial close relatives, and mitochondrial rRNAs representing the four eukaryotic kingdoms (plant, protist, fungi and animal). The two clostridia show minor or unique oligonucleotides in 15 per cent of the families. For the mycoplasmas themselves, this number ranges from a low of 22 per cent to a high of 52 per cent (in *Mycoplasma gallisepticum*). The corresponding numbers for the mitochondria are even more spectacular. Only the plant mitochondrion (which shows 33 per cent minor or unique occurrence oligonucleotides) is as 'normal' as the mycoplasmas; the animal mitochondrion is represented by minor or unique variants in 26 of the 27 families (96 per cent)!

An underlying mechanism

One might question whether the differences between normal bacteria and the mycoplasmas (or mitochondria) merely reflect special functional constraints operating on these ribosomes. However, mycoplasmas appear to produce

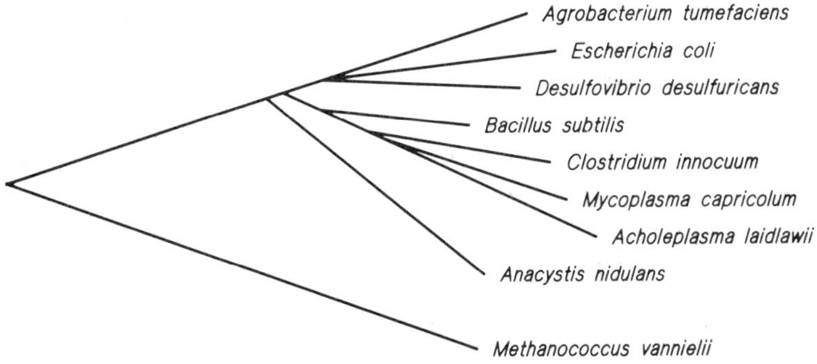

Fig. 7.2 Phylogenetic tree derived from the percentage similarities of Table 7.3 by the procedure of McCarroll *et al.* (1983). The position of the root is arbitrary.

Table 7.3. *Homology matrix for various eubacterial 16S rRNA sequences*

Lower left-hand triangle: Percentage of positions in which a given pair of sequences have the same composition, calculated for those positions represented in all eubacterial sequences only, a total of 1385.

Upper right-hand triangle: Binary association coefficients, S_{AB} values, calculated for the same pairs of sequences (Fox *et al.*, 1977). S_{AB} is the sum of bases in all oligonucleotides (hexamer and larger) common to two oligonucleotide catalogues, divided by half the sum of all bases in all oligonucleotides (hexamer and larger) in the two (Fox *et al.*, 1977).

Abbreviations: *Agrobacterium tumefaciens* – Atume (Yang *et al.*, 1985); *Escherichia coli* – Ecoli (Brosius *et al.*, 1978); *Desulfovibrio desulfuricans* – Desul (Oyaizu & Woese, 1985); *Bacillus subtilis* – Bsubt (Green *et al.*, 1985); *Clostridium innocuum* – Cinno (unpublished); *Mycoplasma capricolum* – Mcapr (Iwami *et al.*, 1984); *Acholeplasma laidlawii* – Alaid (unpublished); *Anacystis nidulans* – Anidu (Tomioka & Sugiura, 1983); and *Methanococcus vannielii* – Mvann (Jarsch & Bock, 1985).

	S_{AB} value								
	Atume	Ecoli	Desul	Bsubt	Cinno	Mcapr	Alaid	Anidu	Mvann
1 Atume	—	29	28	29	24	22	21	24	12
2 Ecoli	80.4	—	35	26	14	17	15	24	08
3 Desul	81.7	80.6	—	27	16	13	12	26	07
4 Bsubt	80.2	79.4	81.3	—	36	31	27	30	05
5 Cinno	76.9	76.5	78.0	83.2	—	36	37	23	05
6 Mcapr	77.7	76.6	76.6	82.5	81.8	—	27	19	04
7 Alaid	76.9	73.9	76.0	79.6	82.2	82.0	—	18	04
8 Anidu	78.3	78.2	79.2	81.2	77.8	76.8	76.7	—	07
9 Mvann	62.5	61.6	62.7	64.0	61.9	61.8	60.9	63.7	—

normal types of protein. The internal environment of the cell is not reported to be unusual in its ionic conditions, optimum temperature, and so on. Mycoplasma ribosomes are sensitive to normal bacterial antibiotics (Razin, 1978). And their closest non-mycoplasma relatives (which share their ribosomal RNA sequence idiosyncracies to some extent – Tables 7.2 and 7.3) appear to the microbiologist as normal bacteria, hence their inclusion in the genus *Clostridium*. Furthermore, the published (Iwami *et al.*, 1984) and unpublished sequences for mycoplasma 16S rRNAs show no idiosyncrasy in secondary structure. Thus, there is little or no reason to consider mycoplasma ribosomes to be anything other than typical bacterial ribosomes functionally. (Whether this is also true for mitochondrial ribosomes is a moot point.) Therefore, the explanation for the unusual patterns shown by mycoplasmas (and perhaps mitochondria) in Table 7.2 does not seem to lie in functional uniqueness of their ribosomes.

The extent to which patterns characteristic of the highly conserved regions in rRNA are broken by the mycoplasmas (and mitochondria) far exceeds what one might expect on the basis of their (rapid) evolutionary rate as measured by overall sequence homology. However, it is reflected in the S_{AB} values of the mycoplasmas, seen in Table 7.3 (for the reason given above).

As discussed above, change in the regions of conserved sequence is rare enough that it would seem to occur by a mechanism involving (nearly) simultaneous mutations at two or more sites. This in turn suggests that the greater variance shown by mycoplasmas in Table 7.2 reflects an increased mutation rate in these lines of descent. While such would increase the general tempo of change in rRNA sequence, its effect on those areas of sequence in which change was a higher-order function of mutation rate would be disproportionately large, and therefore dramatic. And that is what is observed.

A priori there is good reason to suspect that mycoplasma mutation rates are elevated; their genomes are small, one-fourth or less the size of typical eubacterial genomes (Wallace & Morowitz, 1973). If the upper bound to mutation rate in a line of descent is set by the necessity to make less than a critical number of mistakes in replicating the genome, then, other things being equal, a mycoplasma could do this with just as much *overall* accuracy as does *E. coli*, despite having a fourfold higher mutation rate. In addition, there seems to be no reason why the mutation rate in a line of descent would remain as low as those characteristic of normal bacteria if the genome size in that line became reduced. (Increase in the rate is merely a loss of function, so presumably the rate would ultimately rise when not held down by some selective constraint.) Although the mutation rates in mycoplasmas have not been measured, an elevated rate is consistent with the observation that some mycoplasmas lack certain DNA repair capabilities (Ghosh, Das & Maniloff, 1977).

General considerations

The above differences between mycoplasma (and mitochondrial) and normal bacterial rRNAs in patterns of base replacement are indeed striking. If the interpretation of these data is correct, then the microbial (molecular) counterpart of rapid (macro-)evolution has been recognized, and a simple relationship between the tempo and the mode of evolution – at least on the bacterial level – is suggested.

The effect of mutation rate upon the phenotype in a line of descent is far more drastic than is generally appreciated. This is because the significant (drastic, novel, etc.) changes in phenotype tend to involve co-ordinate multiple mutations, and so tend to be higher-order functions of the mutation rate. Put another way, the field of variants associated with a line of descent is strongly affected by the mutation rate in that line. When mutation rate increases, not only do the usual kinds of variants increase in frequency, but the field in a sense becomes dominated by variants that are normally present at negligible levels. This manifests itself in ribosomal RNA in changes in regions of the sequence that tend to be invariant normally. It is manifest in the general phenotype in terms of changes we perceive as novel, bizarre, etc.

Perhaps the best analogy for the ribosomal RNA as an evolutionary chronometer is a clock (counting device) built with multiple hands (counters). The primary or main counter (hand) would be a simple evolutionary distance measuring device, i.e. something that measures average rate × time. However, any other counter, or hand, would be a threshold device, having the property that it tended to register only when the evolutionary (mutation) rate is above a critical level. Such a threshold counter would allow a measurement of peak rates which is independent of average rate and not confused with the usual evolutionary distance measurement.

Ribosomal RNAs measure not only evolutionary tempo (distance), then, but also its mode in a (semi-)independent way. This makes them more powerful chronometers than macromolecules are now considered to be. It also follows that the ribosomal RNA chronometer can delimit the root of phylogenetic trees without having to invoke outgroup species for this purpose. (If the rate of evolution in various groups of organisms is recorded internally, the position of the root of their phylogenetic tree can be constrained.)

It seems likely that the chronometric structure of ribosomal RNA is not unique to that molecule, but is a general characteristic of macromolecules, and indeed this may also be reflected in the general evolutionary characteristics of the whole organism.

Chronic vs episodic rapid evolution and metazoan macroevolution

If the relationship between genome size and mutation rate suggested above is true, the mycoplasmas and mitochondria represent lines that are in chronic states of rapid evolution (so long as their genomes do not enlarge to normal bacterial size). In the metazoan world, however, rapid evolution tends to occur in transient, episodic forms, such as at the beginning of major lines of descent (Simpson, 1944). If the present mechanism were to apply to the episodic form of rapid evolution, additional mechanisms would have to be postulated to account for the increases and decreases of mutation rate that would occur. A discussion of these would not be particularly useful at this time, for the subject is too ill-defined for it to be anything other than an exercise in unbridled speculation. However, conditions that would permit a mutation rate to rise – e.g. periods of rapidly changing environment – and others that would force it to drop again can readily be imagined.

Increased mutation rate would, then, seem a sufficient condition for rapid evolution. Whether it is also a necessary condition has to be resolved by experimentation. It is certainly conceivable that other mechanisms, such as special environmental conditions, could produce phenomena that phenotypically mimic rapid evolution brought about by elevated mutation rate (Mayr, 1942; Simpson, 1944; Wright, 1982). The two mechanisms, however, are easily distinguished: rapid evolution caused by an elevated mutation rate must be accompanied by widespread and novel changes at the molecular level, changes almost all of which are selectively neutral.

A major question is whether, assuming the present mechanism holds for bacterial macroevolution, it then applies to metazoan macroevolution as well. It is counterproductive summarily to dismiss the possibility. If metazoans did not utilize this mechanism as seen on the bacterial level, they might well use a more subtle form of it. To this point we have assumed that elevated mutation rates hold for an organism's entire genome, an assumption which seems reasonable when dealing with bacteria. Eukaryotes, which have much larger genomes, might not survive an episode of elevated mutation rate as readily as bacteria. However, eukaryotes seem to be able to alter locally the mutation rate in the genome (Kreitman, 1983). In such cases, only that aspect of an organism's phenotype controlled by the genes involved would undergo rapid evolution, and the environmental conditions under which this restricted form of rapid evolution would occur might be less drastic than for the global variety. Eukaryotes, moreover, could also employ a different sort of elevated mutation rate, one involving the frequency of recombination, not of point mutation.

REFERENCES

Bonen, L. & Doolittle, W. F. (1976). Partial sequences of 16 S rRNA and the phylogeny of blue-green algae and chloroplasts. *Nature, London*, **261**, 669–73.

Bonen, L. & Doolittle, W. F. (1978). Ribosomal RNA homologies and the evolution of the filamentous blue-green bacteria. *Journal of Molecular Evolution*, **10**, 291–7.

Brooks, B. W., Murray, R. G. E., Johnson, J. L., Stackebrandt, E., Woese, C. R. & Fox, G. E. (1980). Red-pigmented micrococci: a basis for taxonomy. *International Journal of Systematic Bacteriology*, **30**, 627–46.

Brosius, J., Palmer, J. L., Kennedy, J. P. & Noller, H. F. (1978). Complete nucleotide sequence of a 16S ribosomal RNA gene from *Escherichia coli*. *Proceedings of the National Academy of Sciences USA*, **75**, 4801–5.

Eperon, I. C., Anderson, S. & Nierlich, D. P. (1980). Distinctive sequence of human mitochondrial ribosomal RNA genes. *Nature, London*, **286**, 460–7.

Fox, G. E., Peckman, K. R. & Woese, C. R. (1977). Comparative cataloging of 16S ribosomal ribonucleic acid: molecular approach to procaryotic systematics. *International Journal of Systematic Bacteriology*, **27**, 44–57.

Fox, G. E., Stackebrandt, E., Hespell, R. B., Gibson, J., Maniloff, J., Dyer, T. A., Wolfe, R. S., Balch, W. E., Tanner, R., Magrum, L. J., Zablen, L. B., Blakemore, R., Gupta, R., Bonen, L., Lewis, B. J., Stahl, D. L., Luehrsen, K. R., Chen, K. N. & Woese, C. R. (1980). The phylogeny of prokaryotes. *Science*, **209**, 457–63.

Ghosh, A., Das, J. & Maniloff, J. (1977). Lack of repair of ultraviolet light damage in *Mycoplasma gallisepticum*. *Journal of Molecular Biology*, **116**, 337–44.

Gibson, J., Stackebrandt, E. & Woese, C. R. (1985). The phylogeny of the green photosynthetic bacteria: lack of a close relationship between *Chlorobium* and *Chloroflexus*. *Systematic and Applied Microbiology*, **6**, 152–6.

Green, C. J., Stewart, G. C., Hollis, M. A., Vold, B. S. & Bott, K. S. (1985). Nucleotide sequence of *Bacillus subtilis* ribosomal RNA operon, *rrnB*. *Gene*, **37**, 261–6.

Greenwood, P. H. (1979). Macroevolution – myth or reality? *Biological Journal of the Linnean Society*, **12**, 293–304.

Gupta, R., Lanter, J. & Woese, C. R. (1983). Sequence of the 16S ribosomal RNA from *Halobacterium volcanii*, an archaebacterium. *Science*, **221**, 656–9.

Gutell, R. R., Weiser, B., Woese, C. R. & Noller, H. F. (1985). Comparative anatomy of 16S-like ribosomal RNA. *Progress in Nucleic Acid Research and Molecular Biology*, **32**, 155–216.

Hespell, R. B., Paster, B. J., Macke, T. J. & Woese, C. R. (1984). The origin and phylogeny of the bdellovibrios. *Systematic and Applied Microbiology*, **5**, 196–203.

Iwami, M., Muto, A., Yamao, F. & Osawa, S. (1984). Nucleotide sequence of the *rrnB* 16S ribosomal RNA gene from *Mycoplasma capricolum*. *Molecular and General Genetics*, **196**, 317–22.

Jarsch, M. & Bock, A. (1985). Sequence of the 16S ribosomal RNA gene from *Methanococcus vannielii*. *Systematic and Applied Microbiology*, **6**, 54–9.

Kimura, M. (1982). The neutral theory as a basis for understanding the mechanism of evolution and variation at the molecular level. In *Molecular Evolution, Protein Polymorphism and the Neutral Theory*, ed. M. Kimura, pp. 3–58. Tokyo: Japan Scientific Societies Press.

Kimura, M. (1983). *The Neutral Theory of Molecular Evolution*. Cambridge University Press.

Knoll, A. H. & Barghoorn, E. S. (1978). Archean microfossils showing cell division from the Swaziland system of South Africa. *Science*, **198**, 396–8.

Kreitman, M. (1983). Nucleotide polymorphism at the alcohol dehydrogenase locus of *Drosophila melanogaster*. *Nature, London*, **304**, 412–16.

Lowe, D. R. (1980). Stromatolites 3,400-M yr old from the Archaean of western Australia. *Nature, London*, **284**, 441–3.

Ludwig, W., Schleifer, K.-H., Reichenbach, H. & Stackebrandt, E. (1983). A phylogenetic analysis of the myxobacteria *Myxococcus fulvus, Stigmatella aurantiaca, Cystobacter fuscus, Sorangium cellulosum* and *Nannocystis exedens*. *Archives of Microbiology*, **135**, 58–62.

McCarroll, R., Olsen, G. J., Stahl, Y. D., Woese, C. R. & Sogin, M. L. (1983).

Nucleotide sequence of the *Dictyostelium discoideum* small-subunit ribosomal ribonucleic acid inferred from the gene sequence: evolutionary implications. *Biochemistry*, **22,** 5858–68.

Margulis, L. (1970). *Origin of Eucaryotic Cells.* New Haven, Connecticut: Yale University Press.

Mayr, E. (1942). *Systematics and the Origin of Species.* New York: Columbia University Press.

Nomura, M., Mizushima, S., Ozaki, M., Traub, P. & Lowry, P. E. (1969). Structure and function of ribosomes and their molecular components. *Cold Spring Harbor Symposia of Quantitative Biology*, **34,** 49–61.

Oren, A., Paster, B. J. & Woese, C. R. (1984*a*). Haloanaerobiaceae: a new family of moderately halophilic, obligatory anaerobic bacteria. *Systematic and Applied Microbiology*, **5,** 71–80.

Oren, A., Weisburg, W. G., Kessel, M. & Woese, C. R. (1984*b*). *Halobacteroides halobius* gen. nov., sp. nov., a moderately halophilic anaerobic bacterium from the bottom sediments of the Dead Sea. *Systematic and Applied Microbiology*, **5,** 58–70.

Oyaizu, H. & Woese, C. R. (1985). Phylogenetic relationships among the sulfate respiring bacteria, myxobacteria and purple bacteria. *Systematic and Applied Microbiology*, **6,** 157–63.

Paster, B. J., Stackebrandt, E., Hespell, R. B., Hahn, C. M. & Woese, C. R. (1984). The phylogeny of the spirochetes. *Systematic and Applied Microbiology*, **5,** 337–51.

Paster, B. J., Ludwig, W., Weisburg, W. G., Stackebrandt, E., Hespell, R. B., Hahn, C. M., Reichenbach, H., Stetter, K. O. & Woese, C. R. (1985). A phylogenetic grouping of the bacteroides, cytophagas and certain flavobacteria. *Systematic and Applied Microbiology*, **6,** 34–42.

Razin, S. (1978). The mycoplasmas. *Microbiological Reviews*, **42,** 414–70.

Schwartz, R. M. & Dayhoff, M. O. (1978). Origins of prokaryotes, eukaryotes, mitochondria, and chloroplasts. *Science*, **199,** 395–403.

Seilhammer, J. J., Olsen, G. J. & Cummings, D. J. (1984). Paramecium mitochondrial genes: I. Small subunit rRNA gene sequence and microevolution. *Journal of Biological Chemistry*, **259,** 5167–72.

Simpson, G. G. (1944). *Tempo and Mode in Evolution.* New York: Columbia University Press.

Sor, F. & Fukuhara, H. (1980). Séquence nucléotidique du gène de l'ARN ribosomique 15S mitochondrial de la levure. *Compte Rendu de l'Académie des Sciences, Paris*, série D, **291,** 933–6.

Spencer, D. F., Schnare, M. N. & Gray, M. W. (1984). Pronounced structural similarities between the small subunit ribosomal RNA genes of wheat mitochondria and *Escherichia coli. Proceedings of the National Academy of Sciences USA*, **81,** 493–7.

Stackebrandt, E., Ludwig, W., Schubert, W., Klink, F., Schlesner, H., Roggentin, T. & Hirsch, P. (1984). Molecular genetic evidence for early evolutionary origin of budding peptidoglycan-less eubacteria. *Nature, London*, **307,** 735–7.

Stanier, R. Y. (1970). Some aspects of the biology of cells and their possible evolutionary significance. In *Organization and Control in Prokaryotic and Eukaryotic Cells*, 20th Symposium, Society for General Microbiology, ed. H. P. Charles & B. C. J. G. Knight, pp. 1–38. Cambridge University Press.

Tomioka, N. & Sugiura, M. (1983). The complete nucleotide sequence of a 16S ribosomal RNA gene from a blue-green alga, *Anacystis nidulans*. *Molecular and General Genetics*, **191**, 46–50.

Wallace, D. C. & Morowitz, H. J. (1973). Genome size and evolution. *Chromosoma (Berlin)*, **40**, 121–6.

Woese, C. R. (1982). Archaebacteria and cellular origins: an overview. *Zentralblatt für Bakteriologie, Mikrobiologie und Hygiene*, Originale **C3**, 1–17.

Woese, C. R., Debrunner-Vossbrinck, B. A., Oyaizu, H., Stackebrandt, E. & Ludwig, W. (1985*a*). Gram-positive bacteria: possible photosynthetic ancestry? *Science*, **229**, 762–4.

Woese, C. R. & Fox, G. E. (1977). The phylogenetic structure of the procaryotic domain: the primary kingdoms. *Proceedings of the National Academy of Sciences USA*, **74**, 5088–90.

Woese, C. R., Gutell, R. R., Gupta, R. & Noller, H. F. (1983). Detailed analysis of the higher-order structure of 16S-like ribosomal ribonucleic acids. *Microbiological Reviews*, **47**, 621–69.

Woese, C. R., Maniloff, J. & Zablen, L. B. (1980). Phylogenetic analysis of the mycoplasmas. *Proceedings of the National Academy of Sciences USA*, **77**, 494–8.

Woese, C. R. & Olsen, G. J. (1986). Archaebacterial phylogeny: perspectives on the urkingdoms. *Systematic and Applied Microbiology*, **8**, in press.

Woese, C. R., Stackebrandt, E. & Ludwig, W. (1985*b*). What are mycoplasmas: the relationship of tempo and mode in bacterial evolution. *Journal of Molecular Evolution*, **21**, 305–16.

Woese, C. R., Stackebrandt, E., Macke, T. J. & Fox, G. E. (1985*c*). A phylogenetic definition of the major eubacterial taxa. *Systematic and Applied Microbiology*, **6**, 143–51.

Woese, C. R., Stackebrandt, E., Weisburg, W. G., Paster, B. J., Madigan, M. T., Fowler, V. J., Hahn, C. M., Blanz, P., Gupta, R., Nealson, K. H. & Fox, G. E. (1984*a*). The phylogeny of purple bacteria: the alpha subdivision. *Systematic and Applied Microbiology*, **5**, 315–26.

Woese, C. R., Weisburg, W. G., Hahn, C. M., Paster, B. J., Zablen, L. B., Lewis, B. J., Macke, T. J., Ludwig, W. & Stackebrandt, E. (1985*d*). The phylogeny of purple bacteria: the gamma subdivision. *Systematic and Applied Microbiology*, **6**, 25–33.

Woese, C. R., Weisburg, W. G., Paster, B. J., Hahn, C. M., Tanner, R. S., Krieg, N. R., Koops, N.-P., Harms, H. & Stackebrandt, E. (1984b). The phylogeny of purple bacteria: the beta subdivision. *Systematic and Applied Microbiology*, **5**, 327–36.

Wright, S. (1982). The shifting balance theory and macroevolution. *Annual Review of Genetics*, **16**, 1–19.

Yang, D., Oyaizu, Y., Oyaizu, H., Olsen, G. J. & Woese, C. R. (1985). Mitochondrial orgins. *Proceedings of the National Academy of Sciences USA*, **82**, 4443–7.

Zuckerkandl, E. & Pauling, L. (1965). Molecules as documents of evolutionary history. *Journal of Theoretical Biology*, **8**, 357–66.

8 Divergence in inbred strains of mice: a comparison of three different types of data

Walter M. Fitch & William R. Atchley

Introduction

A topic of continuing discussion among evolutionary biologists involves the best method and the best kind of data for reconstructing phylogenetic history. To know whether a result using a particular method or particular data is better or not requires that one know the correct phylogeny. The general problem is that the true phylogeny is rarely known. The exceptions are those instances where humans have observed (in fact, have been responsible for) the splitting of lineages. Such instances include dogs, horses, mice and *Drosophila*.

There is abundant genetic information on the mice and Fitch & Atchley (1985*a*) analysed some of the molecular data for them. Those results will be reviewed here and compared to similar analyses on morphology and life history traits. With several methods and several data sets in hand, we hoped to find at least partial answers to the following two questions. Which methods are best for reconstructing phylogeny? What kinds of data are best for reconstructing phylogeny?

The analyses will show that, from the molecular data, one obtains the correct tree by any one of the five methods tried, but one does not get even close to the correct tree with the morphological data or the life history traits.

Methods and materials

Five methods were used to reconstruct the phylogeny of the mice. These were:

1 the unweighted pair-group method of analysis (UPGMA, Sneath & Sokal, 1973);

2 parsimony (Fitch, 1971);
3 EVOLVES (Fitch & Margoliash, 1967);
4 neighbourliness (Fitch, 1981); and
5 distance Wagner (Farris, 1972).

The 10 strains of mice used in the analyses of the molecular data were (with coat colour) *A/HeJ* (albino), *AKR/J* (albino), *BALB/cJ* (albino), *CBA/J* (agouti), *C3H/HeJ* (agouti), *C57BL/6J* (black), *C57BR/cdJ* (brown), *C58/J* (black), *DBA/1J* (grey) and *DBA/2J* (grey). Known genetic relationships among these inbred strains are shown in Fig. 8.1 according to information in Staats (1980) and Festing (1979).

Molecular data were allelotypes taken from Staats (1980), who records the allele present, where known, for 158 genetic loci from these and many other strains of mice. A list of the 97 loci used can be found in Fitch & Atchley (1985*a*). A summary of those data appears in Table 8.1.

Morphological data were 14 measurements obtained from the mandible of 10-week-old mice (Atchley, Plummer & Riska, 1985*a*, *b*; Atchley & Newman,

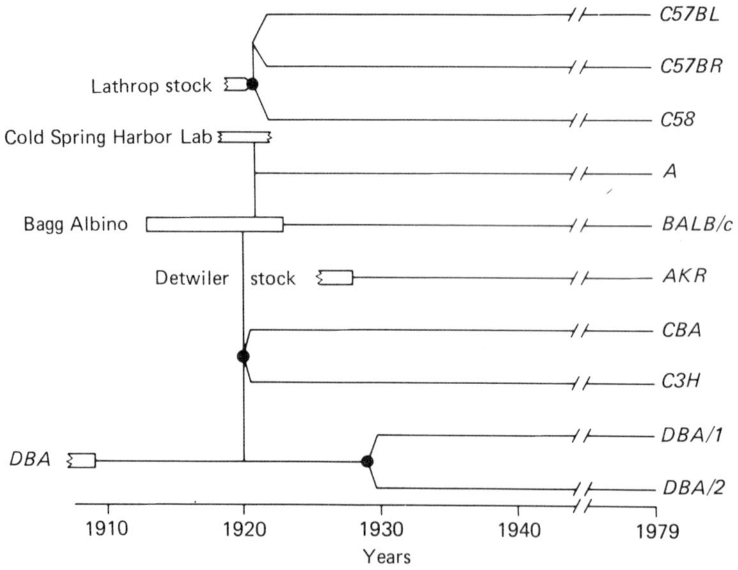

Fig. 8.1 Genealogy of the 10 strains of inbred mice used in this study. Thin lines show periods during which the strains were inbred, open boxes periods of pen breeding. Jagged ends of boxes indicate uncertainty regarding the time point (scale in years at foot). The *CBA* and *C3H* strains were derived from separate brother–sister pairs from the same litter of a Bagg Albino–*DBA* cross. The two *C57* strains were derived from separate brother–sister pairs from one litter. The *C58* strain was derived from another brother–sister pair from a litter sired by the same male as the *C57* strains.

Table 8.1. *Characteristics of the molecular data for 97 examined loci*

Variability	
23	loci unvaried
8	loci with unique variants only
66	loci with more than one non-unique allele
Phenotypic nature	
33	immunological
62	proteins
2	other
Chromosomal location	
87	autosomal
10	unassigned
Missing data	
54 of 970 allelotypes missing	
No strain missing more than 7 allelotypes	
No locus missing more than 2 allelotypes	

1987) using the landmark points shown in Fig. 8.2. These measurements represent an extension of a smaller set analysed by Festing (1973), Lovell & Johnson (1983) and Bailey (1985) in previous studies of mouse mandibular form. Although technically a single bone, the dentary, the mammalian mandible is a developmentally complex structure whose component parts have different embryological origins and may have different controlling factors (Hall, 1978, 1982; Atchley *et al.*, 1985*a*). The genetic correlations between strains among these mandible traits range from 0.01 to unity (Atchley *et al.*,

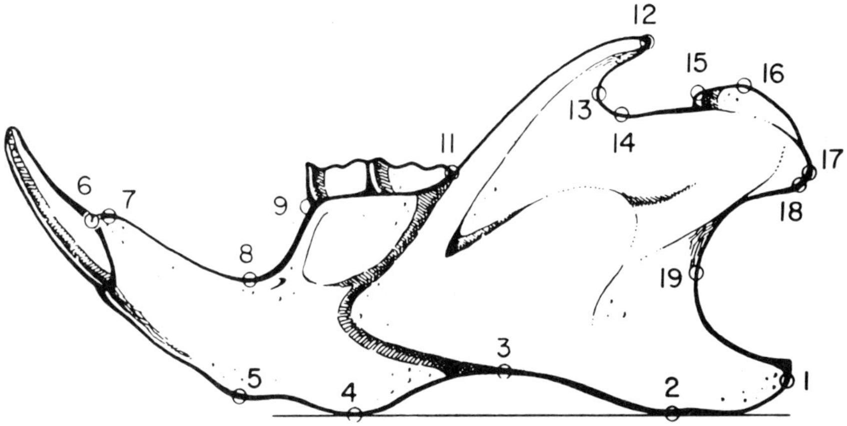

Fig. 8.2 Drawing of a mouse mandible. Numbered points were used as locations to obtain 14 different length measurements to obtain genetic distance based on quantitative morphological traits.

1985*a*). The average genetic correlation among these 14 mandible traits is only 0.32 (±0.02) indicating that, on average, only about 10 per cent of the genetic variability in one mandible dimension (square of the average genetic correlation) can be explained by the genetic covariance (= pleiotropic effects) with a second mandible trait (Atchley *et al.*, 1985*a*).

Among the most commonly measured evolutionary changes in polygenic traits within and between natural populations are changes in body size, rate of growth and reproductive output. Traits relating to these processes are often described as 'life history' attributes and they are among the features most widely studied by ecologists, animal breeders, developmental biologists and geneticists (Dingle & Hegmann, 1982; Lande, 1982; Harvey & Clutton-Brock, 1985; Atchley & Newman, 1987). The several life history traits used for these mouse strains are litter size, body weight at 2 weeks, at weaning, and at 70 days of age, and three growth curve parameters (weight at birth, initial rate of growth, and decline in rate of growth). In random-bred mice of the ICR strain, the genetic correlation between body weight at 2 weeks of age and at weaning (3 weeks of age) is high (around 0.9); however, the genetic correlation between these two traits and 10-week weight is about 0.5 (Riska, Atchley & Rutledge, 1984). The genetic correlation between litter size and 2-week weight is 0.4, while litter size has a genetic correlation of −0.1 with weight at weaning and of 0.3 with 10-week body weight (Atchley & Newman, 1987). Thus, there is considerable genetic independence among several of these life history traits.

Results

Agreement among molecular trees

Table 8.2 records the allelotype differences for the 10 strains of mice. The percentage differences shown were used for reconstructing the genealogies by all tree-building methods except parsimony, which requires the use of the allelotypes directly.

Figure 8.3 shows the results obtained from three analyses of the molecular data. These three trees, and two others obtained by the neighbourliness (Fitch, 1971) and distance Wagner (Farris, 1972) methods, show complete topological agreement with each other in every detail except one. That exception is the order of divergence of *C57BL*, *C57BR* and *C58*. At their origin, the first two strains were formed from two brother–sister pairs from the same litter, and hence were related as full sibs, while *C58* was formed from another brother–sister pair from another litter sired by the same male, and hence is related to the first two as half-sibs. This will affect how different they are from each other, but, as we shall see, not greatly.

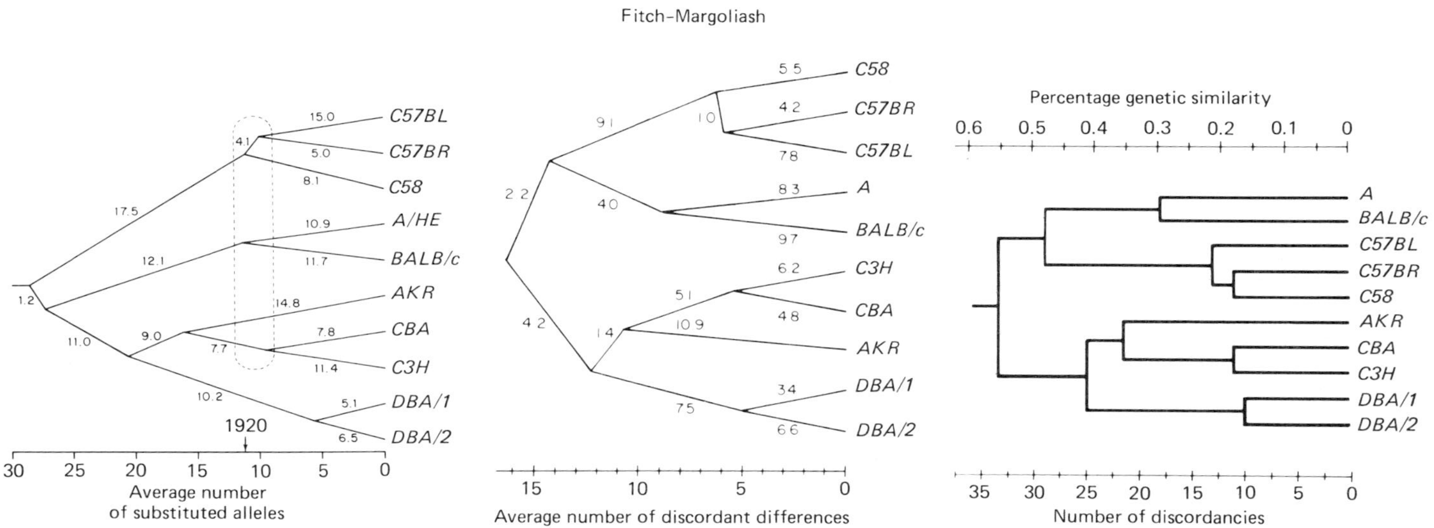

Fig. 8.3 Three trees based on 97 molecular traits. Left: Tree constructed using the parsimony method (Fitch, 1971). Centre: Tree constructed using the Fitch–Margoliash method (Fitch & Margoliash, 1967). Right: Tree constructed using the UPGMA method (Sneath & Sokal, 1973). The centre and right-hand trees were constructed using the data in the upper right half of Table 8.2. All three trees, plus two others not shown, are topologically identical (branching order is the same) to each other and to the genealogy, where known, shown in Fig. 8.1 except for the resolution of the *C57–C58* node (see text for explanation). In the parsimony tree, left, the dotted line encloses nodes of known date, in 1920 and 1921 (cf. Fig. 8.1). The scale of this tree is number of substituted alleles, but if it is converted to a linear time-scale by assigning 1979 to the tips (the date of Festing's book, from which most of the data came) and 1830 (as determined independently) to the root, 1920 comes out at the point marked. This point falls within the dotted outline corresponding to that year, corroborating both the tree and the method used to construct it.

Table 8.2. *Pair-wise differences for 97 loci of 10 inbred mouse strains*

Values below the diagonal are the actual number of allelic differences, while those above the diagonal reflect the percentage of loci with different alleles in those loci for which the allelotypes of both members of the pair are known (note range is 14 per cent – *DBA/1* and */2* – to 54 per cent – *DBA/2* and *C57BL*). The data in the upper right half were used to reconstruct genealogies by all methods except parsimony.

Strain	*C57BL*	*C57BR*	*C58*	*BALB/c*	*A*	*AKR*	*CBA*	*C3H*	*DBA/1*	*DBA/2*
C57BL	—	0.23	0.27	0.41	0.41	0.51	0.51	0.51	0.52	0.54
C57BR	20	—	0.19	0.43	0.44	0.41	0.44	0.49	0.49	0.50
C58	23	15	—	0.47	0.48	0.40	0.44	0.46	0.48	0.50
BALB/c	38	36	39	—	0.24	0.50	0.46	0.41	0.37	0.45
A	39	37	40	22	—	0.43	0.44	0.30	0.38	0.46
AKR	48	35	34	46	40	—	0.31	0.34	0.35	0.39
CBA	47	36	36	41	40	28	—	0.20	0.35	0.31
C3H	47	41	38	37	28	31	18	—	0.38	0.38
DBA/1	49	41	41	34	35	33	32	35	—	0.14
DBA/2	43	36	37	36	37	31	24	30	11	—

Heterozygosity, *h*, is the probability that two gametes chosen at random from the population have the same allele at a particular genetic locus. In wild mice, $h \sim 0.1$. The 10 laboratory strains are not wild and so *h* could, at the time of their origin, be different. In particular, it appears that all 10 strains descended from a cross of a female mouse from central England and a male mouse from eastern Europe or farther east.

This origin is believed because of the following observations. Ferris, Sage & Wilson (1982) and Ferris *et al.* (1983) noticed that all the 'old' strains of inbred mice (our 10 strains are all in this group) contained the same mitochondrial DNA pattern when cut by enzymes that specifically cleave certain nucleotide sequences. That pattern was rare in wild mice. In fact, they found this inbred mouse pattern only in wild mice (*Mus musculus domesticus*) from a small region of central England. Since mitochondria are inherited only from the mother, it may be concluded that the common ancestor of all these 'old' inbred mice included a female from central England.

Bishop *et al.* (1985) discovered a piece of DNA that was found only on the Y (male) chromosome of mice. In examining the distribution of this piece, they discovered that it was specific in wild mice for *Mus musculus musculus*, a subspecies that extends from central Europe eastward. However, among the 10 strains we studied, that piece is found in all six tested by Bishop *et al.* It is therefore a reasonable inference that the common ancestor of all these 'old' inbred mice included a male from somewhere east of central Europe (Fitch & Atchley, 1985*b*).

What effect does this have on our estimate of heterozygosity? Although each of these two ancestors may well have been only 0.1 heterozygous, the progeny of that cross would probably have a higher heterozygosity. Using the data of R. D. Sage (personal communication), we calculated the heterozygosity of a cross between a *M. m. domesticus* mouse from Abingdon, England, and a *M. m. musculus* mouse from Bratislava, Czechoslovakia. This gave an h of 0.21. We will assume this value is representative of the ancestral cross that produced the ancestor of the mice in this study.

How is this related to our sibling problem? We want to know how different one might expect the siblings to be. Note that inbreeding is accomplished by mating a brother with a sister in each generation. The probability that the genetic material for a locus in the egg of a sister and in the sperm of a brother did not derive from the same chromosome of the same parent is 0.75. Multiplied over many generations, the probability that a given genetic locus is not homozygous by descent becomes small (it is about 0.01 after 16 generations). In these studies we are examining lines that have been inbred for over 100 generations. We may, therefore, think of each strain as effectively homozygous at all loci, with the equivalent of a double dose of a gamete. Thus, comparing two inbred lines is like comparing two gametes; that is, the fraction of loci at which they differ might relate to the heterozygosity of the ancestral pool.

But *C57BL*, *C57BR* and *C58* are not two random gametic representatives of the ancestral gene pool which we have estimated above as having a heterozygosity of 0.21. If they were, we would expect them to be different at $0.21 \times 97 = 20$ of the loci tested, irrespective of which two of the three strains we chose to compare. Being all equally different by expectation, we could not expect to distinguish their branching order.

But these three strains are not randomly related. The full sibs (*C57BL* and *C57BR*) would be expected to have 0.25 of their gametic loci identical by descent and hence would show only $0.21 \times 97 \times 0.75 = 16$ differences. Half-sibs would be expected to have only 0.125 of their genetic loci identical by descent, so that half-sibs (*C58*) should differ by $0.21 \times 97 \times 0.875 = 18$ differences.

The distribution of those gametic differences is something of a lottery. One may expect 18 on average, but it could, in fact, easily be anything from 16 to 20 in any particular case, and a similar uncertainty surrounds the expectation of 16 differences. The result is, although the expectations are that the full sibs *C57BL* and *C57BR* should be more closely related to each other than either is to their half-sib *C58*, the variability involved in randomly sampling their gametes suggests that obtaining the desired relationship among these three

Table 8.3. *Morphological divergence of the mandible in seven inbred strains of mice*

The numbers in the lower left half of the table are the genetic divergences based on 14 dimensions of the mandible and are the square root of the Mahalanobis generalized distance. The upper right half shows the molecular distances (from Table 8.2) for comparison, which should be on the basis of their relative, not their absolute, magnitudes.

Strain	*C57BL*	*C58*	*BALB/c*	*A*	*CBA*	*C3H*	*DBA/2*
C57BL	—	0.27	0.41	0.41	0.51	0.51	0.54
C58	4.30	—	0.47	0.48	0.44	0.46	0.50
BALB/c	8.56	6.43	—	0.24	0.46	0.41	0.45
A	8.04	6.71	8.00	—	0.44	0.30	0.46
CBA	6.06	4.74	7.88	6.33	—	0.20	0.31
C3H	8.52	6.43	7.37	6.68	5.96	—	0.38
DBA/2	6.67	4.54	6.59	6.62	3.56	7.31	—

strains would be more accidental than the result of methodological power. It is to this feature that the single exception to congruence among the trees in Fig. 8.3 is most likely due.

Agreement of molecular trees with the known phylogeny

The trees (Fig. 8.3) are also all topologically identical (same branching order) to the known phylogeny with one exception. That exception relates to the fact that the *CBA* and *C3H* strains were formed by a cross between two other strains, *DBA* and *BALB/c*, whose descendants are also in our group. *CBA* and *C3H* are thus hybrids. But none of the standard methods used to reconstruct phylogenies allows for the possibility that two diverging lineages can produce a third by hybridization. Thus, in the trees the *CBA/C3H* pair must be joined to only one lineage, not two. The question is, which one?

Obviously, it should be either the *DBA* or the *BALB/c* lineages, but that still leaves a choice. The members of the litter from which *CBA* and *C3H* derived received equal numbers of chromosomes from each parent, but the *DBA* lineage was already inbred. The result is that all genes received from the *DBA* parent were, or should have been, identical to all the genes at those loci in *DBA* and all of its descendants. The *BALB/c* lineage, on the other hand, was heterozygous, since inbreeding had not yet been started. Thus, the descendants of Bagg albino chromosomes fixed in *CBA* and *C3H* need not be the same as those that went to fixation in the present-day *BALB/c*. It follows that there should be more differences between *CBA/C3H* and *BALB/c* than between *CBA/C3H* and *DBA*, thereby leading to the conclusion that the methods should, if they are doing their job, attach the *CBA/C3H* group closer to *DBA* than to *BALB/c*. As can be seen in Fig. 8.3, all trees show this relation.

The major observation, then, is that all five methods of tree construction yield the same correct, known phylogeny of the 10 inbred strains of mice. This is in rather sharp contrast to the common experience with other large data sets (10 or more taxa), where one often gets a different tree with each new method tried, because of the numerous possibilities. It is undoubtedly a testament to the robustness of the data that the answer is the same with our mice regardless of the method employed. This was in part disheartening to us because the whole project was initially undertaken to test the relative merits of the various tree-making methods and, with these data, all methods proved equally worthy and we were without answer to our first question.

Disagreement of mandible trees with the known phylogeny

Table 8.3 shows, in the lower left half, the genetic divergence for the mandibular data between all pairs of seven strains of inbred mice, all of which were among the ten strains used in the molecular study. Mandible data were not available for the other three strains. The methodology for obtaining these genetic divergence estimates from bone measurements is described by Atchley & Newman (1987). The upper right half of Table 8.3, from the molecular study, contains the fraction of the loci (where both loci were known) that differ between those same pairs. A tree based on the mandibular data is shown on the left of Fig. 8.4.

This tree is very different from those in Fig. 8.3 and from the known genealogy. By virtue of the *DBA* line being already inbred, *CBA* and *C3H* were even more closely related to each other at their origin than were the full sibs, *C57BL* and *C57BR*. Nevertheless, the mandibular data place *C3H* far away from *DBA* and *CBA*. Moreover, since the *CBA/C3H* pair is the result of a cross between *DBA* and the Bagg albino (which became *BALB/c*), it is disconcerting to find *C57BL* and *C58* thrown into the middle of that group.

The pattern of genetic divergence inferred from the molecular and morphological traits is obviously quite different. The Spearman rank order correlation between the molecular and mandibular divergences is only 0.04. Other more robust statistical procedures support the conclusion of unrelatedness of the two sets of data. It is worth pointing out that the mandible is a

Table 8.4. *Divergence of seven life history traits in seven inbred mouse strains*

The numbers in the lower left half of the table are the genetic divergences based on seven life history traits and are the square root of the Mahalanobis generalized distance. The upper right half shows the molecular distances (from Table 8.2) for comparison, which should be on the basis of their relative, not their absolute, magnitude.

Strain	*C57BL*	*C58*	*BALB/c*	*A*	*CBA*	*C3H*	*DBA/2*
C57BL	—	0.27	0.41	0.41	0.51	0.51	0.54
C58	1.26	—	0.47	0.48	0.44	0.46	0.50
BALB/c	1.89	1.96	—	0.24	0.46	0.41	0.45
A	1.01	1.45	2.19	—	0.44	0.30	0.46
CBA	1.06	1.07	1.64	1.37	—	0.20	0.31
C3H	2.09	1.90	1.66	2.37	2.16	—	0.38
DBA/2	2.37	3.21	3.40	2.30	2.45	4.18	—

structure frequently used in inferring evolutionary relationships from fossil data (Moore, 1981). Therefore, it is surprising that the morphological divergences in the mandible are so poorly correlated with the known genealogical relationships of the mice. Our results also suggest that Festing's (1973) observations, that the amount of divergence in the mandible among substrains of *C57BL* mice is linear with respect to time, should not be extrapolated to larger time intervals, even intervals as brief as those in our study.

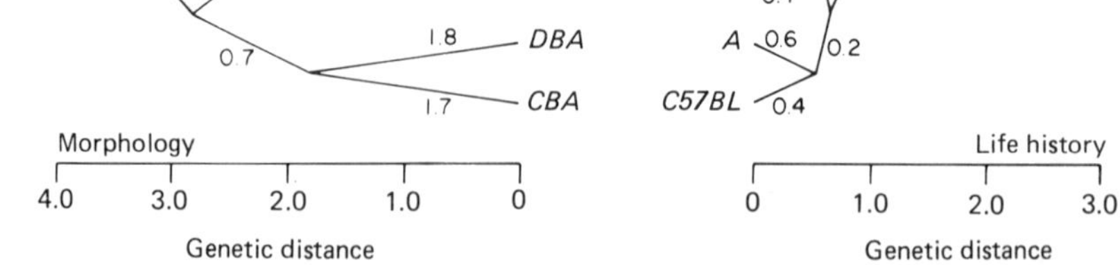

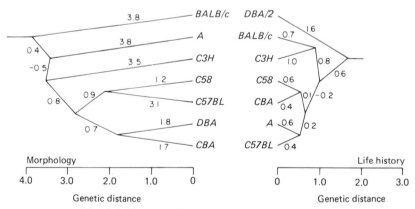

Fig. 8.4 Two trees based on mandibular and life history traits. Left: Tree obtained using the UPGMA method on the mandibular data in the lower left of Table 8.3. Right: Tree obtained using UPGMA on the life history data in the lower left of Table 8.4.

Disagreement of life history traits with the known phylogeny

The lower left half of Table 8.4 shows the genetic divergence for the life history traits between all pairs of the same seven strains of mice used in the mandible study. Again, the molecular data are shown in the upper right half. A tree based on the life history traits is shown on the right of Fig. 8.4.

This tree is very different from those in Fig. 8.3 and from the known genealogy. *BALB/c* is far removed from *A*, *CBA* is far removed from *C3H* and both are far removed from *DBA*, and *C57BL* and *C58* are each made the sister-group of other taxa.

Again, as in the case of the morphological data, the pattern of genetic divergence inferred from the molecular and life history traits is quite different. The Spearman rank order correlation between the molecular and life history divergences is −0.07.

Discussion

The perfect topological agreement between the trees based on molecular data and the known history of the mouse strains is remarkable. The results with the mandible data and the life history traits suggest an answer to our second question, 'what kinds of data should be used for reconstructing evolutionary history?' With severe caveats (below), molecular data appear to be superior to morphological data and to life history data.

Now for those severe caveats.

1. The number of traits we sampled for the morphology (14) and life history (7) is too few and the traits themselves too intercorrelated. Even though the mandible characters can be shown to be largely independent of each other, suites of well-chosen characters that are as disparate as possible (non-bony as well as bony) are less likely to be subject to errors of a sort not reflective of genealogy. We would probably obtain similarly divergent results for the molecular data if we subsampled the 97 loci, work that we propose to do.
2. A question that needs to be asked is whether evolutionary patterns are best reflected by structural gene changes or regulatory gene changes. The monogenic (i.e. allelic) molecular data reflect differences in structural genes, whereas the mandibular and life history traits are polygenic and presumably include regulatory loci as well.

3 We have performed this analysis for only one known genealogy. Until a similar analysis is performed with similar results on other organisms, the generality of our result is suspect.
4 The time span of divergence in this study (70 years) is nearly infinitesimal compared to the times encompassed by most evolutionary questions. The changes seen in morphology in 70 years might be little more than random fluctuations with respect to the changes seen over palaeontological time.
5 The taxa are somewhat atypical in that they have been manipulated by man. However, the evolutionary processes (selection, random fixation of alleles, drift, etc.) must be the same as those that act in natural populations.
6 The divergence seen here in the molecular, morphological and life history traits is not the result of fixing new mutations but of having segregated the heterozygosity resulting from a cross made about 150 years ago (Fitch & Atchley, 1985*a*, *b*).
7 No cladistic techniques were employed on the mandible and life history traits (because the data, measurements modified by removal of environmental effects, are not amenable to cladistic treatment).
8 The morphological traits employed here are quantitative measurements (lengths, widths, etc.), whereas most attempts at reconstructing phylogenics from morphology use largely qualitative traits (presence, absence, etc. of homologous features).

The above caveats do not make us doubtful about the results obtained for these strains of mice; they simply make us wary of trying to extrapolate from them.

Our efforts to answer the two initial questions concerning what is the best method for phylogeny reconstruction and what are the best data to use in such methods have given us surprising results but no solid answers. But the prospects of additional work giving better answers are good.

The first surprise was that all methods gave the same correct phylogeny for the molecular data. They may not do that, however, for smaller subsets of the 97 loci. If it were to occur that the UPGMA method stopped getting the correct tree after the data set was reduced to (say) 60 loci while the parsimony method continued to get the correct answer until the data set was reduced to 30 loci, one might reasonably conclude that the latter method was less sensitive to sampling errors than the former.

The second surprise was the very low correlation among the molecular, morphological and life history data sets. That might not remain true for larger sets of morphological and life history data.

We are pursuing the two principal questions by reducing the size of the molecular data set and increasing the size of the the other data sets.

ACKNOWLEDGEMENTS

This work was supported by National Science Foundation grants BSR-84-00682 to WMF and DEB-8109904 to WRA.

REFERENCES

Atchley, W. R. & Newman, S. (1987). The genetic basis of morphological divergence: I, divergence in mandible form among inbred mouse strains. *Genetics*, in press.

Atchley, W. R., Plummer, A. A. & Riska, B. (1985*a*). Genetics of mandible form in the mouse. *Genetics*, **111**, 555–77.

Atchley, W. R., Plummer, A. A. & Riska, B. (1985*b*). A genetic analysis of size-scaling patterns in the mouse mandible. *Genetics*, **111**, 579–95.

Bailey, D. W. (1985). Genes that affect the shape of the murine mandible. *Journal of Heredity*, **76**, 107–14.

Bishop, C. E., Boursot, P., Baron, B., Bonbomme, F. & Hatat, D. (1985). Most classical *Mus musculus domesticus* laboratory mouse strains carry a *Mus musculus musculus* Y chromosome. *Nature, London*, **315**, 70–2.

Dingle, H. & Hegmann, J., eds. (1982). *Evolution and Genetics of Life Histories*. New York: Springer-Verlag.

Farris, J. S. (1972). Estimating phylogenetic trees from distance matrices. *American Naturalist*, **106**, 645–68.

Ferris, S. D., Sage, R. D. & Wilson, A. C. (1982). Evidence from mtDNA sequences that common laboratory strains of inbred mice are descended from a single female. *Nature, London*, **295**, 163–5.

Ferris, S. D., Sage, R. D., Prager, E. M., Ritte, U. & Wilson, A. C. (1983). Mitochondrial DNA evolution in mice. *Genetics*, **105**, 681–721.

Festing, M. (1973). A multivariate analysis of subline divergence in the shape of the mandible in C57BL/Gr mice. *Genetical Research*, **21**, 121–32.

Festing, M. (1979). *Inbred Strains in Biomedical Research*. Oxford University Press.

Fitch, W. M. (1971). Toward defining the course of evolution: minimum change for a specific tree topology. *Systematic Zoology*, **20**, 406–16.

Fitch, W. M. (1981). A non-sequential method for constructing a hierarchical classification. *Journal of Molecular Evolution*, **18**, 30–7.

Fitch, W. M. & Atchley, W. R. (1985*a*). Evolution in inbred strains of mice appears rapid. *Science*, **228**, 1169–75.

Fitch, W. M. & Atchley, W. R. (1985*b*). Rapid mutations in mice? *Science*, **230**, 1408–9.

Fitch, W. M. & Margoliash, E. (1967). The construction of phylogenetic trees: a generally applicable method utilizing estimates of the mutation distance obtained from cytochrome c sequences. *Science*, **155**, 279–84.

Hall, B. K. (1978). *Developmental and Cellular Skeletal Biology*. New York: Academic Press.

Hall, B. K. (1982). How is mandibular growth controlled during development and evolution? *Journal of Craniofacial Genetics*, **2,** 45–9.

Harvey, P. H. & Clutton-Brock, T. H. (1985). Life history variation in primates. *Evolution*, **39,** 559–81.

Lande, R. (1982). A quantitative genetic theory of life history evolution. *Ecology*, **63,** 607–15.

Lovell, D. P. & Johnson, F. M. (1983). Quantitative genetic variation in the skeleton of the mouse. I. Variation between inbred lines. *Genetical Research*, **42,** 169–82.

Moore, W. J. (1981). *The Mammalian Skull.* Cambridge University Press.

Riska, B., Atchley, W. R. & Rutledge, J. J. (1984). A genetic analysis of targeted growth in mice. *Genetics*, **107,** 79–101.

Sneath, P. H. A. & Sokal, R. R. (1973). *Numerical Taxonomy*. San Francisco: W. H. Freeman & Co.

Staats, J. (1980). Standardized nomenclature for inbred strains of mice: Seventh listing. *Cancer Research*, **40,** 2083–128.

INDEX

Entries in **bold** denote principal references or definitions; entries in *italic* refer to illustrations.